INTRODUCTION TO

UNMANNED AIRCRAFT SYSTEMS

INTRODUCTION TO UNMANNED AIRCRAFT SYSTEMS

Edited by
Richard K. Barnhart • Stephen B. Hottman
Douglas M. Marshall • Eric Shappee

CRC Press is an imprint of the
Taylor & Francis Group, an **informa** business

CRC Press
Taylor & Francis Group
6000 Broken Sound Parkway NW, Suite 300
Boca Raton, FL 33487-2742

International Standard Book Number: 978-1-4398-3520-3 (Hardback)

Library of Congress Cataloging-in-Publication Data

Introduction to unmanned aircraft systems / [edited by] Richard K. Barnhart ... [et al.].
-- 1st ed.
p. cm.
"A CRC title."
Includes bibliographical references and index.
ISBN 978-1-4398-3520-3 (hardcover : alk. paper)
1. Drone aircraft. 2. Drone aircraft--Automatic control. I. Barnhart, Richard K. II. Title: Unmanned aircraft systems.

UG1242.D7I587 2012
623.74'69--dc23 2011036198

Visit the Taylor & Francis Web site at
http://www.taylorandfrancis.com

and the CRC Press Web site at
http://www.crcpress.com

Contents

Preface

It is our pleasure to present this first edition of *Introduction to Unmanned Aircraft Systems*. It is well understood that the unmanned aircraft system (UAS) industry is highly dynamic and constantly evolving with the advancement of science and technological enablement. As such, the aim of this text is to identify and survey the fundamentals of UAS operations, which will serve as either a basic orientation to UAS or as a foundation for further study. This contributed work targets introductory collegiate courses in UASs and was birthed out of an unsuccessful search for suitable texts for such a course. The chapters have been individually contributed by some of the nation's foremost experts in UAS operations at the collegiate level; therefore, the reader may note some variation in writing style. It was decided to leave the contributions in this form in the interest of preserving the author's intent, thereby improving the quality of information contained herein. This text is written from a nonengineering, civilian, operational perspective aimed at those who will operate or employ UASs for a variety of future missions.

This publication would not have been possible without the close cooperation of all the editors and contributors; a heartfelt thank you to all who gave of your time to make this possible.

Your feedback is welcomed as a basis for future editions of this text as the industry continues to advance.

Acknowledgments

The editors would like to personally thank all who contributed to this work and their families for allowing them the time to make this sacrifice. The end result is worth it. We would also like to thank University of North Dakota graduate student Kristen Steuver who helped find our mistakes and ensure that we said what we meant to say. For all your services we are eternally grateful.

Editors

Richard K. Barnhart, Ph.D., is professor and head of the Aviation Department at Kansas State University (K-State; Salina) in addition to serving as the executive director of the Applied Aviation Research Center at K-State, which oversees the newly established Unmanned Aerial Systems program office. Dr. Barnhart is a member of the graduate faculty at K-State and holds a commercial pilot certificate with instrument, multiengine, seaplane, and glider ratings. He also is a certified flight instructor with instrument and multiengine ratings. Dr. Barnhart holds an airframe and power-plant certificate with inspection authorization. In addition, he holds an A.S. in aviation maintenance technology from Vincennes University (Indiana), a B.S. in aviation administration from Purdue University (West Lafayette, Indiana), an MBAA from Embry-Riddle Aeronautical University (Daytona Beach, Florida), and a Ph.D. in educational administration from Indiana State University (Terra Haute). Dr. Barnhart's research agenda is focused in aviation psychology and human factors. His industry experience includes work as a research and development (R&D) inspector with Rolls-Royce Engine Company, and systems instructor for American Trans-Air airlines. Most recently, Dr. Barnhart was associate professor and acting department chair of the Aerospace Technology Department at Indiana State University, where he was responsible for teaching flight and upper-division administrative classes. Courses taught include Aviation Risk Analysis, Citation II Ground School, King Air 200 Flight, Air Navigation, Air Transportation, Instrument Ground School, and many others.

Stephen B. Hottman is the director of the Unmanned Aircraft System (UAS) Technical Analysis and Applications Center (TAAC) and associate dean for Research and Development and senior deputy director of the New Mexico State University (NMSU; Las Cruces) Physical Science Laboratory (PSL). The TAAC includes all of the UAS research, development, test, and evaluation (RDT&E) and operations that are taking place in the National Airspace System (NAS). The UAS TAAC is evolving into a center of influence for global UAS RDT&E and airspace integration and access. Hottman's areas of responsibility include leadership for two separate departments. The 21st Century Aerospace Department includes all of the unmanned aircraft work performed since 1999; and missile, rocket, and telemetry support to the Missile Defense Agency, Space and Missile Defense Command, National Aeronautics and Space Administration (NASA), and other agencies. The Columbia Scientific Balloon Facility (Department) is a government-owned contractor-operated NASA carrier program providing all operations, R&D, and engineering support for near space activity (120–160K AGL) with several ton payloads operating for over a month at a time. Launch operations occur at numerous international sites utilizing balloons up to 60 M ft2.

Prior to NMSU, Hottman worked in the industry and focused primarily on optimizing the effectiveness of the human in manned weapon systems. His last industry position in 1998 was leading a human systems strategic business unit focused on human performance, chem/bio, and training that was the largest provider of such services to the Department of Defense (DoD), other federal agencies, and international sponsors. His career has included an eclectic variety of domains, including directed energy weapons development; chemical biological warfare vulnerability and countermeasures for ground and aircrew members; command, control, and communication; a variety of aviation programs; and training primarily in support of the war fighter.

Hottman is a Ph.D. candidate in psychology (engineering psychology) at New Mexico State University, has an M.S. in industrial engineering (human factors specialization), and a B.S. in psychology; both degrees are from Texas A&M University. He has more than 100 publications and presentations. He is a life member of the National Defense Industrial Association (NDIA); and the Aerospace Medical Association, a member of the Association for Unmanned Vehicle Systems International (AUVSI), Human Factors and Ergonomics Society, and a member of the Southwest Border Security Task Force.

Douglas M. Marshall, J.D., is division manager, UAS Regulations & Standards Development at the Physical Science Laboratory, New Mexico State University, where he is involved in UAS research and a variety of efforts concerned with the safe integration of unmanned aircraft systems into national and international airspace. He previously held the position of professor of aviation at the University of North Dakota, where he directed the aviation master's program and taught courses in aviation law, airline labor law and policy, aviation economics, airline management, organizational behavior and management theory, and aviation regulations and policy. Marshall has been engaged full-time on UAS-related activities for more than 6 years, is the coeditor of two books on aviation topics, and is the author of numerous published articles on aviation law, regulations, and remotely piloted aircraft.

Marshall currently serves on the Radio Technical Commission for Aeronautics (RTCA) SC-203, American Society of Testing and Materials (ASTM) F-38, and Society of Automotive Engineers (SAE) G-10 committees, the Association for Unmanned Vehicle Systems International (AUVSI) Advocacy Committee, the Arctic Monitoring and Assessment Program's UAS Expert Group, and The Arctic Research Interest Group. He previously served on the Federal Aviation Administration's (FAA) Small UAS Aviation Rulemaking Committee, whose recommendations are under consideration by the FAA. He has also served on the Steering Committee, Civil Applications of Unmanned Aerial Systems Conference (Boulder, Colorado); the Department of Defense's Joint Integrated Product Team for UAS; and several other committees dedicated to the development of UASs and has delivered presentations on international aviation regulations and airspace issues at conferences around the world.

In his previous career, Marshall practiced law for three decades, specializing in maritime and aviation litigation and employment law, representing several major airlines in labor- and employment-related matters, and was the general counsel and later president of a commuter airline. He holds a private pilot's certificate with commercial and instrument ratings.

Marshall earned a bachelor of arts degree at the University of California, Santa Barbara, in 1969, and a juris doctor from the University of California's Hastings College of the Law in 1972. He is an active member of the American Institute of Aeronautics and Astronautics (AIAA), where he is vice chair of the Legal Aspects of Aeronautics Technical Committee, as well as a member of the American Academy for the Advancement of Science, the Association for Unmanned Vehicle Systems International, and regularly serves on panels and delivers presentations on UAS-related topics.

Eric Shappee serves as an associate professor of aviation at Kansas State University at Salina in the professional pilot program. He teaches numerous aviation courses, which include: Introduction to Aviation, System Safety, Safety Management, and Introduction to Unmanned Aerial Systems. Shappee holds a commercial pilot certificate with instrument, multiengine, and glider ratings. He is also a certified flight instructor with gold seal. Shappee holds two associate degrees from Antelope Valley College (Palmdale, California), a bachelors in aeronautical science and a masters in aeronautical science and safety from Embry-Riddle Aeronautical University. Shappee's main area of focus in aviation is safety. He has developed several risk assessment tools for K-State and other aviation organizations. Further, he was recently named as a member of the International Society of Air Safety Investigators, a membership earned through professional service and experience. Shappee has been active in the field of aviation since 1986, and teaching since 1995. During his career in aviation, Shappee has also spent time working with unmanned aerial systems including the Predator and Aerosonde.

Contributors

Ted Beneigh, Ph.D., serves on the faculty at Embry-Riddle Aeronautical University, where for the past 34 years he worked as a flight instructor, check airman, and director of flight standards, in addition to most recently serving for 12 years in the Aeronautical Science Department as a tenured professor. Dr. Beneigh holds an associate's degree in aeronautical studies, a bachelor of science in aeronautical science, a masters of science in aeronautical science, and a Ph.D. in international business administration. In addition to his academic qualifications, Dr. Beneigh holds an Airline Transport Pilot Certificate with more than 16,000 accumulated flight hours. He currently serves as the program coordinator for the newly established UAS degree.

Josh Brungardt has served as the chief pilot for High Performance Aircraft Training, EFIS Training, and Lancair Aircraft. He currently is serving as the UAS director for Kansas State University's Aviation Department. In 2010, Brungardt attended senior officer training on the Predator UAS at Creech Air Force Base with the 11th Reconnaissance SQ. In addition to completing over 100 first flights on experimental aircraft he has served as a test pilot and instructor pilot to the U.S. Air Force and other defense contractors. Brungardt also specializes in electronic flight information systems (EFIS) and avionics integration. In 2006, Brungardt started a pilot training company called EFIS Training, which focuses on pilots transitioning to glass cockpits. He holds ATP and CFII ratings with over 4000 hours, as well as having raced at the Reno National Air Races. Brungardt received a bachelor's degree in airway science and an associate's degree in professional pilot from Kansas State University at Salina.

Igor Dolgov, Ph.D., attained a bachelor of science and engineering (B.S.E.) in computer science from Princeton University (New Jersey), with a certificate in intelligent systems and robotics. He continued his education at Arizona State University (Tempe) where he received the University Graduate Scholar award while earning a Ph.D. in psychology as part of the Cognition, Action, & Perception Program in the Department of Psychology, and serving as an National Science Foundation Integrated Graduate Education and Research Traineeship Program (NSF IGERT) fellow with the School of Arts, Media, & Engineering. Dolgov currently holds a tenure-track position in the Engineering Psychology Program at New Mexico State University (NMSU) and has established the PACMANE laboratory, which investigates Perception, Action, and Cognition in Mediated, Artificial, and Naturalistic Environments. He also collaborates extensively with NMSU's Cognitive Engineering Laboratory as part of the Physical Science Laboratory's (PSL) 21st Century Aerospace Program. His transdisciplinary-applied research is focused on the perceptual and cognitive capabilities of UAS operators as well as the efficacy of current UAS interfaces. Additionally, the laboratory is evaluating the impact of varying perceived task importance in the trust of automated systems, as well as general graphic user interface design principles. Along with Dolgov's applied activities, theoretical research at the PACMANE Lab spans across ecological psychology and embodied cognition paradigms, with a focus on the interplay between human perception and action, in both naturalistic and artificial contexts.

Lisa Jo Elliott, Ph.D., received a B.A. in psychology from the University of Arizona (Tucson) in 2004, an M.A. in experimental psychology from New Mexico State University in 2006, and a Ph.D. in engineering psychology from NMSU in 2011. She has worked as a human factors engineer for IBM, as an engineering psychologist for the Department of Defense, and as a researcher at the NMSU Physical Science Laboratory/UAS Flight Test Center. Dr. Elliott is currently an assistant professor of psychology at the University of South Florida–Polytechnic in Lakeland, Florida.

Charles "Chuck" Jarnot, USAR LTC (Ret.), is a graduate of the Aviation Department of Western Michigan and holds a master's degree in aeronautical science from Embry-Riddle Aeronautical University. Jarnot spent his career as a U.S. Army rotary wing aviator and aviation advisor serving in numerous locations throughout the United States and the world, including Korea and Afghanistan, where he most recently served in a UAS support role helping to field the MMist CQ-10A Snowgoose®. Jarnot has also served as an adjunct faculty member with Kansas State University teaching an introductory UAS course. In addition, he has field experience on other UAS platforms including Insitu's ScanEagle® where he is fielding this asset in his current position in the Middle East. One of Jarnot's passions is aviation history and he has a particular interest in rotary wing history.

Matthew J. Rambert is an engineering psychology graduate student and researcher at New Mexico State University, with experience in human factors, computer repair, system analysis, usability and usage analysis, and experience running a research laboratory.

Jeremy D. Schwark is an engineering psychology graduate student and researcher in the Cognitive Engineering Laboratory at New Mexico State University. His research focuses on automation, human–computer interaction, and interface design.

Caitriana Steele is a postdoctoral research associate with the Jornada Experimental Range at New Mexico State University. She is an earth observation scientist with specific interests in the use of remote sensing and geospatial technologies for natural resource management.

Bryan Stewart lives in Las Cruces, New Mexico. He received a B.S. in electrical engineering and an M.S. in electrical engineering from New Mexico State University, Las Cruces, in 2004 and 2006, respectively. Stewart currently works at White Sands Missile Range (WSMR) in the Systems Engineering Directorate as part of the Networks and Controls division. He works with the unmanned autonomous systems test focus area from the Test Resource Management Center. Some of his duties at WSMR are to provide support for target control and network communication in addition to developing automated testing procedures.

Prior to working at WSMR, Stewart worked in the Electromagnetic Group at the Physical Science Laboratory (PSL) designing and fabricating patch antennas. He also did work as part of a Defense Threat Reduction Agency (DTRA)-funded effort to examine mitigation techniques against high-power microwaves (HPMs) for commercial and industrial buildings. Stewart is working to obtain a Ph.D. in electrical engineering.

Glen Witt was a consultant on Airspace Operations for the Technical Analysis and Applications Center at New Mexico State University (NMSU). He retired in 2009. Witt began his work career in the U.S. Air Force in 1955. After completing the Air Force's air traffic control (ATC) course at Keesler Air Force Base, he received a year of special ATC training with the Civil Aeronautics Administration (CAA), the Federal Aviation Administration (FAA) predecessor, at the San Antonio, Texas, Air Route Traffic Control Center (ARTCC). At the completion of this training, he was assigned to Lakenheath Radar Air Traffic Control Center, Lakenheath Air Force Base, England, where he performed air traffic controller and supervisor duties for the remainder of his military career.

In 1960, Witt began his employment with the FAA at the Albuquerque ARTCC and worked at that location until his retirement in 1994. Witt's work experience within the FAA included air traffic controller, first- and second-level supervisor, and for the last 10 years of his career, manager of the facility's Airspace and Procedures Office. He received numerous achievement awards and outstanding performance ratings during his tenure with the FAA. During the later stages of his FAA career, he was acknowledged as being one of the FAA's leading authorities on airspace matters and en route ATC procedures. Following his retirement from the FAA in 1994, Witt was employed as a consultant by the Defense Evaluation Support Activity (DESA) and the U.S. Air Force in support of unmanned aerial vehicle (UAV) flights in the National Airspace System (NAS). His achievement in obtaining the FAA's authorization for the Department of Defense (DoD) to operate the Predator UAV (medium/altitude endurance) in NAS, other than within restricted area or warning area airspace and without the use of chase aircraft, is considered a milestone by those knowledgeable of UAV operations. Since 1995, Witt has also provided similar consulting services for NASA's Environmental Research Aircraft and Sensor Technology (ERAST) program. His excellent knowledge of the operation of the ATC system, ATC procedures, airspace management, and appropriate Federal Aviation Regulations places him in a distinct position of being able to assist appropriate authorities to develop UAV operational criteria and standards.

1 History

Charles Jarnot

CONTENTS

1.1 THE BEGINNING

The history of unmanned aircraft is actually the history of all aircraft. From centuries past when Chinese kites graced the skies to the first hot air balloon, unmanned flying craft came first before the risk of someone climbing on board occurred. One early user of unmanned aircraft was by the Chinese General Zhuge Liang (180–234 AD) who used paper balloons fitted with oil-burning lamps to heat the air; he then flew these over the enemy at night to make them think there was a divine force at work. In modern times, unmanned aircraft have come to mean an autonomous or remotely piloted air vehicle that flies about mimicking the maneuvers of a manned or human-piloted craft. Even the name assigned to unoccupied aircraft has changed over the years as viewed by aircraft manufacturers, civil aviation authorities, and the military. Aerial torpedoes, radio controlled, remotely piloted, remote control, autonomous control, pilotless vehicle, unmanned aerial vehicles (UAVs), and drone are but some of the names used to describe a flying machine absent of humans.

In the early years of aviation, the idea of flying an aircraft with no one inside had the obvious advantage of removing the risk to life and limb of these highly experimental contraptions. The German aviation pioneer Otto Lilienthal, circa 1890s, employed unmanned gliders as experimental test beds for main lifting wing designs and the development of lightweight aero structures. As a result, several mishaps are recorded where advances were made without injury to an onboard pilot. Although such approaches to remove people from the equation were used, the lack of a satisfactory method to affect control limited the use of these early unmanned aircraft. Early aviation developmental efforts quickly turned to the use of the first "test pilots" to fly these groundbreaking craft. Further advances beyond unmanned gliders proved painful as even pioneer Lilienthal was killed flying an experimental glider in 1896.

As seen in the modern use of unmanned aircraft, historically unmanned aircraft often followed a consistent operational pattern, described today as the three D's, which stand for *dangerous*, *dirty*, and *dull*. *Dangerous* being that someone is either trying to bring down the aircraft or where the life of the pilot may be at undue risk operationally. *Dirty* is where the environment may be contaminated by chemical, biological, or radiological hazards precluding human exposure. Finally, *dull* is where the task requires long hours in the air making manned flight fatiguing, stressful, and therefore not desirable.

1.2 THE NEED FOR EFFECTIVE CONTROL

The Wright Brothers' success in flying the first airplane is more of a technical success story in solving the ability to control a piloted, heavier-than-air craft. Doctor Langley, the heavily government-financed early airplane designer competing with those two bicycle mechanics from Ohio, also wrestled with the problem of how to control an airplane in flight. Doctor Langley's attempts with a far more sophisticated and better powered airplane ended up headfirst in New York's harbor, not once, but twice over the issue of flight control. After the Wright Brothers taught the fledgling aviation world the secrets of controlled flight, namely, their wing-warpin approach to roll control, development experienced a burst of technical advancement, furthered by the tragedy of World War I. The demands of the 1914–1918 war on early aviation drove an incredible cycle of innovations in all aspects of aircraft design ranging from power plants, fuselage structures, lifting wing configurations, to control surface arrangements. It was in the crucible of the war to end all wars that aviation came of age and, along with this wave of technological advancement came the critical but little recognized necessity of achieving effective flight control.

1.3 THE RADIO AND THE AUTOPILOT

As is often the case with many game-changing technological advances, inventions of seemingly unrelated items combined in new arrangements to serve as the catalyst for new concepts. Such is the case with unmanned aircraft. Even before the first Wright Brothers' flight in 1903, the famous electrical inventor Nicola Tesla promoted the idea of a remotely piloted aircraft in the late 1890s as a flying guided bomb. His concept appears to have been an outgrowth of his work building the world's first guided underwater torpedoes called the "telautomation" in 1898. Tesla preceded the invention of the radio in 1893 by demonstrating one of the first practical applications of a device known as a full spectrum spark-gap transmitter. Tesla went on to help develop frequency separation and is attributed by many as the real inventor of the modern radio.

While the electrical genius Tesla was busy designing the first electric architecture of the City of New York, another inventor, Elmer Sperry, the founder of the famed flight control firm that today bears his name, was developing the first practical gyrocontrol system. Sperry's work, like Tesla's, focused initially on underwater torpedoes for the Navy. He developed a three-axis mechanical gyroscope system that took inputs from the gyros and converted them to simple magnetic signals, which in turn were used to affect actuators. The slow speed of water travel and weight not being as critical an issue for seacraft, allowed Sperry to perfect his design of the world's first practical mechanical autopilot. Next, Sperry turned his attention to the growing new aircraft industry as a possible market for his maritime invention, not for the purpose of operating an aircraft unmanned, but as a safety device to help tame early unstable manned aircraft, and to assist the pilot in maintaining their bearings in bad weather. Sperry began adopting his system of control on early aircraft with the help of airframe designer Glenn Curtis. Together they made a perfect team of flyer–designer and automation inventor. Following excellent prewar progress on the idea, the demand during World War I to find new weapons to combat Germany's battleships combined the inventions of the radio, airplane, and mechanical autopilot to field the world's first practical unmanned aircraft, an aerial torpedo.

1.4 AERIAL TORPEDO: THE FIRST MODERN UNMANNED AIRCRAFT (MARCH 6, 1918)

In late 1916, with war raging in Europe, the U.S. Navy, a military arm of a still neutral country, funded Sperry to begin developing an unmanned aerial torpedo. Elmer Sperry put together a team to tackle the most daunting aerospace endeavor of the time. The Navy contract directed Sperry to build a small, lightweight airplane that could be self-launched without a pilot, fly unmanned out to 1000 yards guided to a target and detonate its warhead at a point close enough to be effective against a warship. (See Figure 1.1.) Considering that the airplane had just been invented 13 years earlier, the ability to even build an airframe capable of carrying a large warhead, against an armored ship, a sizable radio with batteries, heavy electrical actuators, and a large mechanical three-axis gyrostabilization unit was by itself incredible, but then integrating these primitive technologies into an effective flight profile—spectacular.

Sperry tapped his son Lawrence to lead the flight testing conducted on Long Island, New York. As the United States entered the World War I in mid-1917, these various technologies were brought together to begin testing. It is a credit to the

FIGURE 1.1 Early unmanned aircraft. (Photo courtesy USAF Museum.)

substantial funding provided by the U.S. Navy that the project was able to weather a long series of setbacks, crashes and outright failures of the various pieces that were to make up the Curtis N-9 Aerial Torpedo. Everything that could go wrong did. Catapults failed; engines died; airframe after airframe crashed in stalls, rollovers, and crosswind shifts. The Sperry team persevered and finally on March 6, 1918, the Curtis prototype successfully launched unmanned, flew its 1000-yard course in stable flight and dived on its target at the intended time and place, recovered, and landed, and thus the world's first true "drone." Thus, the unmanned aircraft system was born.

Not to be outdone by the Navy, the Army invested in an aerial bomb concept similar to the aerial torpedo. This effort continued to leverage Sperry's mechanical gyrostabilization technology and ran nearly concurrent with the Navy program. Charles Kettering designed a lightweight biplane that incorporated aerostability features not emphasized on manned aircraft such as exaggerated main wing dihedral, which increases an airplane's roll stability, at the price of complexity and some loss in maneuverability. The Ford Motor Company was tapped to design a new lightweight V-4 engine that developed 41 horsepower weighing 151 pounds. The landing gear had a very wide stance to reduce ground roll over on landings. To further reduce cost and to highlight the disposable nature of the craft, the frame incorporated pasteboard and paper skin alongside traditional cloth. The craft employed a catapult system with a nonadjustable full throttle setting.

The Kettering aerial bomb, dubbed the *Bug*, demonstrated impressive distance and altitude performance, having flown some tests at 100 miles distance and 10,000-ft altitudes. To prove the validity of the airframe components, a model was built with a manned cockpit so that a test pilot could fly the aircraft. Unlike the Navy aerial torpedo, which was never put in service production, the aerial bomb was the first mass-produced unmanned aircraft. While too late to see combat in World War I, the aircraft served in testing roles for some 12 to 18 months after the war. The aerial bomb had a supporter in the form of then Colonel Henry "Hap" Arnold, who later became a five-star general in charge of the entire U.S. Army Air Forces in World War II. The program garnered significant attention when Secretary of War Newton Baker observed a test flight in October 1918. After the war, some 12 Bugs were used alongside several aerial torpedoes for continued test flights at Calstrom Field in Florida.

1.5 THE TARGET DRONE

Surprisingly, most of the world's aviation efforts in unmanned aircraft after World War I did not pursue weapon platforms like the wartime aerial torpedo and bomb. Instead, work focused primarily on employing unmanned aircraft technology as target drones. In the interwar years (1919–1939) the aircraft's ability to influence the outcome of ground and naval warfare was recognized, and militaries around the world invested more in antiaircraft weaponry. This in turn created a demand for realistic targets and the unmanned target drone was born. Target drones also played a key role in testing air war doctrine. The British Royal Air Force was in a debate with the Royal Navy over the ability of an airplane to sink a ship. In the early 1920s, General Billy Mitchell of the Army Air Corps sunk a war prize German battleship and subsequent older target warships to the dismay of the U.S. Navy. The counterargument to these acts was that a fully manned ship armed with antiaircraft guns would easily shoot down attacking aircraft. The British used unmanned target drones flown over such armed warships to test the validity of the argument. In 1933, to the surprise of all, a target drone flew over 40 missions above Royal Navy warships armed with the latest antiaircraft guns without being shot down. Unmanned aircraft technology played a key role in formulating air power doctrine and provided key data that contributed to America, England, and Japan concluding that aircraft carriers, which played such vital a role in upcoming World War II, were a good investment.

In the United States, the target drone effort was influenced by the development of the Sperry Messenger, a lightweight biplane built in both manned and unmanned versions as a courier for military applications and as a possible torpedo carrier. Some 20 of these aircraft were ordered. The U.S. Army identified this class of aircraft in 1920 as a Messenger Aerial Torpedo (MAT). This effort waned in the early 1920s and Sperry Aircraft Corporation withdrew from active unmanned aircraft design with the untimely death of Lawrence Sperry, the son of the founder and victim of an aircraft accident.

As the U.S. Army lost interest in the MAT program, the service turned its attention to target drones. By 1933, Reginald Denny had perfected a radio-controlled airplane only 10 ft long and powered by a single-cylinder 8 hp engine. The Army designated this craft the OQ-19, also known later as the MQM-33. Some 48,000 of these nimble light craft were produced and they served throughout the World War II as the world's most popular target drone.

In the late 1930s, the U.S. Navy returned to the unmanned aircraft arena with the development of the Navy Research Lab N2C-2 Target Drone (Figure 1.2). This 2500-pound radial engine biplane was instrumental in identifying the deficiencies in Naval antiaircraft prowess. Much like the earlier Royal Air Force experience with the Royal Navy where drones survived numerous passes on well-armed

FIGURE 1.2 Curtis N2C-2 drone. (Photo courtesy of U.S. Navy.)

warships, the U.S. Navy battleship *Utah* failed to shoot down any N2C-2 drones that were making mock attacks on the ship. Curiously enough, the U.S. Navy added yet another title by describing this class of unmanned drones as NOLO (No Live Operator Onboard). The Navy target drone program of the late 1930s developed the technique of a manned aircraft controlling an unmanned aircraft in flight, which was rediscovered and used to great effect in Iraq in 2007.

During the same interwar years the British Royal Navy attempted to develop an unmanned aerial torpedo and an unmanned target drone both utilizing the same fuselage. Several attempts were made to launch these craft from ships with little success. The Royal Aircraft Establishment (RAE) finally gained a measure of success with a Long Range Gun with Lynx engine called together as the *Larynx*. This program was followed by the Royal Air Force automating an existing manned aircraft as its first practical target drone. This effort involved utilizing the Fairey Scout 111F manned aircraft converted as a gyrostabilized radio plane, now referred to as the *Queen*. Of the five built, all crashed on their initial flight save the last one, which proved successful in gunnery sea trials. The next evolution was to take the Fairey flight control system with the excellent and highly stable DeHavilland Gypsy Moth, now called the *Queen Bee*. This proved much more reliable than the earlier Queen, and the Royal Air Force placed an order for 420 target drone Fairey Queen Bees. This led to the designation of an unmanned aircraft being described by the letter *Q* to denote unmanned operation. This protocol was adopted by the U.S. Military. Although unverified by research, the term *drone* is believed by some to have originated with the *Queen* name as meaning "a bee or drone."

During the interwar years almost all nations with an aviation industry embarked on some form of unmanned aircraft. These efforts were mainly in the form of target drones. Germany was an exception with the work of inventor Paul Schmidt who pioneered the pulse jet as a low-cost, simple, high-performance thrusting device in 1935. His work was reviewed by Luftwaffe General Erhard Milch who recommended the new pulse jet be adapted to unmanned aircraft.

1.6 WWII U.S. NAVY ASSAULT DRONE

The U.S. Navy leveraged its experience with the 1930s N2C-2 Target Drone, which was controlled by a nearby manned aircraft in flight, to develop a large-scale aerial torpedo now renamed as an assault drone. Initially, the assault drone effort took the form of the TDN-1 built in a 200-unit production run in early 1940. This aircraft had a wingspan of 48 ft and was powered by twin six-cylinder O-435 Lycoming engines with 220 hp each in a high wing configuration (Figure 1.3). The aircraft was intended to be employed as a bomb or torpedo carrier in high-risk environments to mitigate the risk to aircrews. The groundbreaking advance in this unmanned aircraft was the first use of a detection sensor in the form of a primitive 75-pound RCA television camera in the nose to provide a remote pilot better terminal guidance from standoff distances. Given the relatively poor reliability and resolution of the first TVs, this was indeed a remarkable feat of new technology integration. The TDN-1 was superseded by a more advanced model called the Navy/Interstate TDR-1 Assault Drone. Some 140 examples were built and a Special Air Task Force (SATFOR) was organized and sent to the Pacific Theater and used against the Japanese in the Bougainville Island Campaign in 1944 with limited but definable success. Operationally, a Navy Avenger Torpedo Bomber was flown as the guiding aircraft equipped with radio transmitters to affect radio control. A television receiver was installed on the Avenger for an operator onboard to guide the drone to its target from as much as 25 miles away. About 50 aircraft were employed against various targets such as gun emplacements with about a 33% success rate.

FIGURE 1.3 TND-1. (Photo courtesy of U.S. Navy.)

The U.S. Navy and Army Air Forces then turned to outfitting older, four-engine bombers into unmanned aircraft in the European Theater against Nazi Germany to destroy high-priority targets such as V-1 Buzz Bomb launch sites. This led to the first example of pitting unmanned aircraft to destroy other unmanned aircraft. Operation Anvil, as it was called, consisted of stripping out Navy PB4Y Privateers (the Navy version of the Consolidated B-24) and packing over 10 tons of high explosives along with a Sperry-designed, three-axis autopilot; radio control links; and an RCA television in the nose. The aircraft were over their normal gross weight and utilized manned pilots to take off, then once in cruise flight, the pilots would bail out over friendly England while the aircraft was controlled by a nearby manned bomber and guided to its target. Operations commenced in August 1944, with disastrous results. On the first mission, the aircraft blew up shortly after takeoff, killing Navy Lieutenants Wilford J. Wiley and Joseph P. Kennedy. The latter was President John F. Kennedy's older brother and son of the former U.S. Ambassador to England, Joseph Kennedy. Numerous failures in equipment canceled the program along with the rapid advancement of Allied forces in Europe negating the reason for the concept.

1.7 WWII GERMAN V-1 BUZZ BOMB

The most significant unmanned aircraft of the World War II was Nazi Germany's V-1 Buzz Bomb (Vengeance Weapon-1). Based on the earlier 1930s work by inventor Paul Schmidt in developing a practical pulse jet, the aircraft integrated an advanced, lightweight, and reliable three-axis gyrostabilized autopilot, a radio signal baseline system for accurate launch point data, and a robust steel fuselage that was resistant to battle damage. The V-1 represented the first mass-produced, cruise-missile-type unmanned aircraft, and its configuration influenced many postwar follow-on unmanned aircraft designs (Figure 1.4).

The V-1 was manufactured by Fieseler Aircraft Company in large numbers with over 25,000 built. This high number makes the V-1 the most numerous combat

FIGURE 1.4 German V-1.

unmanned aircraft in history excluding modern hand-launched platforms. The aircraft was flexible in being both ground and air-launched. It utilized a powerful pneumatic catapult system, which is a familiar feature on many modern-day unmanned aircraft systems. The pulse jet was a simple lightweight, high-thrust device that operated on the principle of cycle compression—explosions at about 50 times a second employing closing veins to direct the gas toward the exhaust. These cycles created the hallmark "buzz" sound made by the engines in flight. Although not fuel efficient by traditional jet engine standards, the pulse jet was cheap to produce, provided high thrust, was reliable and could operate with significant battle damage. The V-1 was also the world's first jet-powered unmanned aircraft weighing about 5000 pounds with an impressive 1800-pound warhead.

Operationally the V-1 was primarily employed from ground-launch rail systems. A small number were air launched from Heinkel 111 bombers, making the V-1 the world's first air-launched unmanned aircraft as well. Some 10,000 V-1s were launched against Allied cities and military targets killing some 7000 people. About 25% were successful and when compared to its fairly low cost, the V-1 was an effective unmanned aircraft, massed produced, and employing many firsts for autonomous-flown aircraft. It influenced future designs and provided the historical pretext to fund many more sophisticated unmanned programs during the follow-on Cold War. The U.S. Navy built a reverse-engineered copy for use in the invasion of Japan and launched improved versions from submarines on the surface, gaining yet another title as the world's first naval-launched, jet-powered, unmanned cruise missile.

1.8 WWII GERMAN MISTLETOE

The teaming of manned and unmanned aircraft was not the exclusive domain of the Allies in World War II. The Germans, in addition to the V-1, built a significant number of piggyback aircraft configurations known as Mistletoe Bombers. The main issue with the effectiveness of the V-1 was that it was not very accurate in flying to its desired end point. Mistletoe was an attempt to deal with this problem by temporarily combining a manned aircraft to guide the unmanned aircraft through a large portion of its flight profile, which would separate near the termination point and guide the unmanned aircraft to the target. About 250 such examples were built normally involving the mating of a JU-88 and a Me-109 fighter FW-190. The concept had little success mainly do to operational challenges and not technical problems.

The German Mistletoe concept could be termed more of a guided bomb than an unmanned aircraft, and several gliding guided bombs were developed by the Germans with some success. The lines between guided missile and unmanned aircraft are not always clear, and in World War II, the V-1 assault drones, explosive-packed, radio-controlled bombers, and the piggyback Mistletoe configuration all involved forms of an airplane, which places them in the category of unmanned aircraft. This distinction is far less clear in the view of future cruise missiles, which are more closely related to their ballistic cousins than airplanes.

1.9 EARLY UNMANNED RECONNAISSANCE AIRCRAFT

As we have seen from the beginning of the first successful unmanned aircraft flight in 1918 through to the World War II, unmanned aircraft have been employed mainly in the target drone and weapon delivery roles. Unmanned aircraft development in the follow-on Cold War years shifted dramatically toward reconnaissance and decoy missions. This trend has continued today where nearly 90% of unmanned aircraft are involved in some form of data gathering in the military, law enforcement, and environmental monitoring arenas. The main reason why unmanned aircraft were not employed in World War II for reconnaissance had more to do with the imagery technology and navigation requirements than the aircraft platforms themselves. Cameras in the 1940s required relatively accurate navigation to gain the desired areas of interest and navigation technology of the day could not compete as well as a trained pilot with a map. This changed in the postwar years with the advent of radar mapping, better radio navigation, Loran-type networks, and inertia navigation systems all enabling an unmanned aircraft to fly autonomously to and from the target area with sufficient accuracy.

One of the first reconnaissance high-performance unmanned aircraft to be evaluated was the Radio Plane YQ-1B high-altitude target drone modified to carry cameras, subsequently GAM-67. This turbojet-powered aircraft was primarily air launched from B-47 aircraft and were proposed to be used in the suppression of enemy antiaircraft destruction (SEAD) role. Cameras were also proposed but the program was canceled after only about 20 were built. Poor range and high cost were given as the reasons for cancellation.

1.10 RADAR DECOYS: 1950s–1970s

The Vietnam War of the 1960s and early 1970s created a high demand for countermeasures to Soviet-built surface-to-air missiles (SAMs) used by the North Vietnamese. The missile threat relied extensively on radar detection of American aircraft. Jamming these radars was attempted with mixed results. However, even under the best circumstances, jamming ground-based radars with airborne systems was problematic in that the ground system probably has access to more power, enabling the radar to overcome the jamming emitter. A more effective solution is to fool the radar into believing it has locked on to a real aircraft and having it waste its expensive missiles on a false target. The U.S. Air Force embarked on such a solution by developing a series of unmanned aircraft to decoy enemy SAM batteries.

To fool a radar signal into believing a decoy resembles an American B-52 Bomber for example, the aircraft does not need to be built physically to resemble the real aircraft. Only minor radar reflectors are needed to create a return radar signal that mimics the actual bomber. As a result the unmanned Air Force decoys were small in size but had the desired effect. The most numerous example of a radar decoy was the McDonnell ADM-20 Quail, which could be carried inside the bomb bay of a B-52 and air launched prior to the bombing run. The Quail was about 1000 pounds, had a range of 400 miles, and could mimic the speed and maneuvers of a B-52. As radar resolution improved, the decoys became less effective and most were out of service by the 1970s.

1.11 LONG-RANGE RECON UNMANNED AIRCRAFT SYSTEMS: 1960s–1970s

The U.S. Air Force pioneered the first mass-produced, long-range, high-speed unmanned aircraft designed to conduct primarily reconnaissance missions but evolved into a wide array of tasks from suppression of enemy air defenses to weapons delivery. The Ryan model 147, later renamed the AQM-34 Lightning Bug and Firefly series, has the longest service record for an unmanned aircraft. Designed as an initiative of the Ryan Aircraft Company in the late 1950s from an earlier target drone, the aircraft was powered by a turbojet, employed low drag wing and fuselage configuration and could reach altitudes in excess of 50,000 ft and speeds of 600 knots (high subsonic).

The Bug, as it was called by its operators, had a long career and flew in a wide range of high-and low-altitude profiles performing electronic signal-gathering intelligence, camera reconnaissance, and various decoy radar signal transmissions. A frequent violator of communist airspace, many were shot down, but enough successfully completed their missions to justify their use. The aircraft underwent many modifications over its operational use spanning the early 1960s to 2003. Many unique and groundbreaking technologies were employed in the Bug unmanned aircraft including air launched from the wing store of modified DC-130 aircraft to midair parachute snag recovery from H-2 Jolly Green helicopters. The AQM-34, as it was renamed late in its career, performed high-priority missions of great national importance such as reconnaissance missions during the 1960s Cuban Missile Crisis, to relatively mundane tasks as a target drone for fighter aircraft air-to-air missiles (Figure 1.5).

FIGURE 1.5 AQM-34. (Courtesy of U.S. Air Force.)

1.12 FIRST HELICOPTER UNMANNED AIRCRAFT SYSTEMS: 1960s–1970s

The U.S. Navy's QH-50 DASH, fielded in the early 1960s, established several firsts for unnamed aircraft. This unusual stacked, counterrotating rotary wing aircraft was the first unmanned helicopter and the first unmanned aircraft to take off and land on a ship at sea. The requirement for the Drone Anti-Submarine Helicopter (DASH) was to extend the delivery range of antisubmarine homing torpedoes. A typical Destroyer in the early 1960s could detect a submarine to ranges over 20 miles but could only launch weapons at less than 5 miles. This small, compact, unmanned helicopter only needed to fly off to the maximum detection range and drop its homing torpedoes over the submerged submarine. The QH-50 DASH used remote control via a pilot on the fantail of a ship to take off and land, then employed a gyrostabilizer autopilot to direct the craft to a location that was tracked by the launching ship's radar. Over 700 were built and were used from 1960 to the mid-1970s where they finished up their career as towing targets for antiaircraft gunnery. Several countries operated the aircraft including France and Japan.

1.13 THE HUNT FOR AUTONOMOUS OPERATION

From the very first unmanned aircraft, designers strived to gain as much independent flight operation from manned ground control as possible. Military requirements called for maximum standoff distance, long endurance, and significant data streams from onboard sensors. The demand for data further competed with bandwidth for flight control transmission further driving the need for self-flight or autonomous operation. Enemy jamming may delay sensor transmission but disrupting required flight control information may cause the loss of the aircraft. Cognizant of the British ability to jam its signals, the German V-1 Flying Bomb of World War II intentionally employed a crude, fully autonomous flight control and navigation system based on mechanical gyros, timers, and some primitive preprogramming involving fuel shutoff to initiate the termination dive. It was not until the advent of small, lightweight digital computers, inertia navigation technology, and finally the global positioning system (GPS) satellite network that autonomous unmanned aircraft operation gained flight autonomy on par with a human-piloted vehicle.

Lightweight computer technology developed in the 1970s, which led to the worldwide explosion in personal computers and the digitalization of everyday items from wristwatches to kitchen appliances, played the most significant role in unmanned aircraft autonomy. With each advance in computing power and cashe memory retrieval, unmanned aircraft gained greater flexibility in addressing changes in winds and weather conditions as well as new variables affecting the mission equipment payloads. Mapping data could now be stored aboard the aircraft, which not only improved navigation but also enabled more accurate sensor camera imagery.

1.14 THE BIRTH OF THE TWIN BOOM PUSHERS

The U.S. Marine Corps' groundbreaking work in the late 1960s with the Bikini built by Republic, laid the foundation of what was to become the most popular UAS configuration, leading to today's RQ-7 Shadow, which is the most numerous UAS outside of the hand-launched Raven. The Marine Corps Bikini fuselage focused on providing the camera payload with a nearly unobstructed field of view attained by placement in the nose section. This led to a pusher engine arrangement further simplified by a twin boom tail. Although delta pushers were attempted, most notably the Aquila UAS, this aerodynamic configuration made weight and balance a more challenging proposition since the elevator moment arms were generally fixed, whereas the twin booms could be easily extended.

In the late 1970s, capitalizing on the Marine Corps Bikini configuration, the Israelis developed a small tactical battlefield surveillance UAS called the Scout, built by Israel Aerospace Industries (IAI). The Scout was accompanied by IAI Decoy UAV-A and the Ryan-built Mabat. The decoys were designed to be flown against SAM batteries so as to fool their radar into activating early or even firing a missile on the drone itself. The Mabat was designed to collect antiaircraft radar signals associated with SAM batteries. Finally, the Scout was designed to exploit the actions of the other two in order to put eyes on the SAM batteries for targeting information and damage assessment after a strike. In addition, the Scout provided close-up battlefield imagery to maneuvering ground commanders, a first for unmanned aircraft. This approach differed greatly from all the previous reconnaissance UAS platforms in that their imagery was more operational and strategic with film being developed afterward or even electronically transmitted to a collection center for analysis. The advances in small-sized computers enabled this real-time bird's-eye view to a maneuvering leader on the ground to directly influencing the decision process on small groups of soldiers or even individual tank movement.

Israeli forces made significant advances in battlefield situational awareness during the June 1982 Bekaa Valley conflict between Israeli and Syrian forces. Operation Peace for Galilee, as it was called by Israel, involved an Israeli ground offensive against Hezbollah terrorists occupying southern Lebanon. Syria, allied with Hezbollah, occupied a large portion of the Bekaa Valley with a sizable ground force consisting of large numbers of new Soviet tanks and heavy artillery. Syrian forces were supported by sophisticated Soviet-built SAM batteries. Israel used a combination of jet-powered decoys and Mabat signal-gathering UAS to detect and identify the Syria SAM battery operating frequencies, and then using the Scout with other manned assets quickly destroyed most of the SAM threat enabling the Israeli ground forces to maneuver with close air support. The Scout UAS, with its twin boom pusher configuration, flew along the sand dunes of the Bekaa Valley and identified Syrian tanks with near real-time data feed to maneuvering Israeli small-unit commanders. This eye-in-the-sky advantage enabled a smaller force to move with greater speed, provided excellent targeting data to Cobra attack helicopters, and directed very effective artillery fires. The Scout UAS was too small to be picked up and tracked by Syrian Soviet-designed radar and proved to be too difficult to observe by fast-flying Syrian jet fighters. The

1982 Bekaa Valley experience initiated a worldwide race to develop close-battle unmanned aircraft.

1.15 DESERT STORM: 1991

Whereas the short 1982 Israeli–Syrian Bekaa Valley campaign represented the first use of close battle UAS, Desert Storm in 1991 represented the first wide-scale employment. The United States and its allies used unmanned aircraft continuously from Desert Shield through Desert Storm. The most numerous system employed was the now-familiar twin boom pusher configuration of the Pointer and Pioneer (Figure 1.6). The aircraft was a joint Israeli–U.S. effort that used a snowmobile 27 hp engine, flew via a remote control joystick on the ground, had a range of about 100 miles, and required an altitude of 2000 feet to maintain a line-of-sight transmission data link. Fully autonomous flight was technically possible but these aircraft opted to have a manned pilot remotely operating the aircraft to achieve more responsive battlefield maneuvering at a desired point of interest. GPS and computer power were not yet sufficiently integrated to enable ground operators to simply designate waypoints on short notice. Also, imagery feeds via satellite links were not sufficiently developed at that small size to affect transmission of data. During Desert Storm, U.S. forces flew some 500 UAS sorties. The Pointer and Pioneers guided artillery, even directing the heavy 16-inch gunfire from the battleship *Iowa*. There is a documented case where a group of Iraqi soldiers attempted to surrender to a Pointer flying low over the desert.

Most militaries around the world concluded after the Desert Storm experience that UAS platforms did indeed have a role to play in spotting enemy locations and directing artillery fires. Conversely, most military analysts concluded that the vulnerable data links precluded UAS use across the board as a replacement for many manned aircraft missions and roles. This opinion was based in part on the limitations of the line-of-sight data link of the Pointer and Pioneer, and a deep-rooted

FIGURE 1.6 Pioneer. (Courtesy of U.S. Navy.)

cultural opposition by manned aircraft pilots and their leadership. A large segment of a nation's defense budget is dedicated to the procurement of military aircraft and the training and employment of large numbers of pilots, navigators and other crew members. Most air forces choose their senior leaders after years of having proved themselves in the cockpit flying tactical aircraft. The very idea of cheaper, unmanned aircraft replacing manned platforms ran against what President Eisenhower warned as the self-fulfilling "military–industrial complex."

1.16 OVERCOMING THE MANNED PILOT BIAS

From the 1990s to the terrorist attacks of 9/11, unmanned aircraft made slow progress, leveraging the increases in small, compact, low-cost computers and the miniaturization of a more accurate GPS signal. However, the barrier to widespread acceptance lay with manned aircraft platforms and the pilots who saw UAS technology as replacing their livelihoods. When 9/11 struck, the U.S. Army had only 30 unmanned aircraft. In 2010, that number was over 2000. The argument against unmanned aircraft had finally given way to the low cost, the reduced risk, and the practicality of a drone, as the press still calls them today, performing the long, boring missions of countless hours of surveillance in both Iraq and Afghanistan. With a person still in the loop of any lethal missile leaving the rails of an Air Force Predator UAS, the "responsibility" argument has for the time being been addressed.

1.17 WILL UNMANNED AIRCRAFT SYSTEMS REPLACE MANNED AIRCRAFT?

The band of unmanned aircraft control runs from a completely autonomous flight control system independent of any outside signals to one that employs a constant data link enabling a pilot to remotely fly the aircraft and, of course, variations in between. A fully autonomous aircraft could in theory fly without the effects of enemy signal jamming and carry out a variety of complex missions. The disadvantage is that a fully autonomous flight control system can be simulated in a computer, enabling the enemy to develop counters to the system much in the same way as video gamers do with autonomous opponents. Once the program flaws are identified, it becomes a simple task to defeat the autonomous system. Additionally, fully autonomous systems will most likely not be allowed to employ lethal force since the chain of responsibility is nonexistent. At the other end of the spectrum, an aircraft that depends on an outside signal, no matter how well it is encrypted, has the potential to be jammed or worse: directed by the enemy through a false coded message. Even if true artificial intelligence is developed enabling an unmanned aircraft to act autonomously with the intuitiveness of a human being, the responsibility factor will prevent UAS from fully replacing manned aircraft. This is even truer with civil applications of passenger travel where at least one "conductor" on board will be required to be held accountable for the actions of the aircraft and to exercise authority over the passengers.

DISCUSSION QUESTIONS

1.1 List and discuss the three D's of UAS employment.
1.2 What is considered to be the first modern unmanned aircraft and in what year did it make its first successful flight?
1.3 Discuss the groundbreaking advances with the WWII U.S. Navy assault drone.
1.4 What was the most significant unmanned aircraft of WWII?
1.5 Discuss the various uses of unmanned aircraft from 1918 to today.

2 Unmanned Aircraft System Elements

Joshua Brungardt

CONTENTS

2.1 INTRODUCTION

2.1.1 What Makes Up an Unmanned Aircraft System (UAS)

In this chapter we will briefly discuss the elements that combine to create a UAS. Most civilian unmanned systems consist of an unmanned or remotely piloted aircraft, the human element, payload, control elements, and data link communication architecture. A military UAS may also include elements such as a weapons system platform and the supported soldiers. Figure 2.1 illustrates a common UAS and how the various elements are combined to create the system.

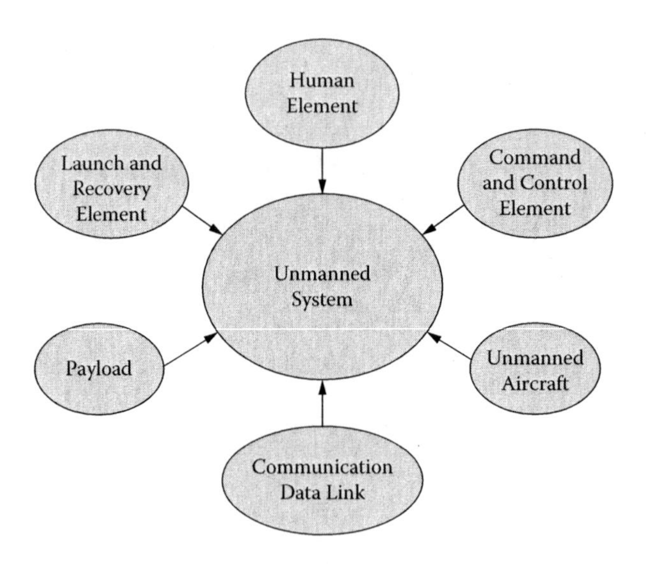

FIGURE 2.1 Elements of an unmanned aircraft system (UAS).

2.2 REMOTELY PILOTED AIRCRAFT

Unmanned aircraft are fixed-wing, rotor-wing, or lighter-than-air vehicles that fly without a human on board. In more recent years there has been a push to change the term *unmanned aircraft* (UA) to remotely piloted aircraft (RPA) or remotely piloted vehicle (RPV). Unmanned aircraft is really a misnomer considering how much human involvement is crucial to the operation of the system.

RPAs are categorized into five groups by the U.S. Department of Defense as seen in the following table.

UAS Category	Max Gross Takeoff Weight	Normal Operating Altitude (ft)	Airspeed
Group 1	<20 pounds	<1200 above ground level (AGL)	<100 knots
Group 2	21–55 pounds	<3500 AGL	< 250 knots
Group 3	<1320 pounds	<18,000 mean sea level (MSL)	
Group 4	>1320 pounds		Any airspeed
Group 5		>18,000 MSL	

Note: If a UAS has even one characteristic of the next higher level, it is classified in that level.

2.2.1 FIXED-WING

A fixed-wing UAS has many missions including intelligence gathering, surveillance, and reconnaissance, or ISR. Some military fixed-wing UAS have adapted a joint mission combining ISR and weapons delivery, such as the General Atomics Predator series of aircraft. The Predator™ was originally designed for an ISR mission with an aircraft designation of RQ-1. In the military aircraft classification system the *R* stands for reconnaissance and the *Q* classifies it as an unmanned aerial system. In recent years however the Predator's designation has been changed to MQ-1, the *M* standing for multirole, having recently been used to deliver hellfire missiles.

Fixed-wing UAS platforms have the advantage of offering operators long flight duration for either maximizing time on station or maximizing range. Northrop Grumman's RQ-4 Global Hawk™ has completed flights of more than 30 hours covering more than 8200 nautical miles. Fixed-wing platforms also offer the ability to conduct flights at much higher altitudes where the vehicle is not visible with the naked eye.

The disadvantages of fixed-wing UAS platforms are that the logistics required for launch and recovery (L&R) can be very substantial (known as a large logistical "footprint"). Some may require runways to land and takeoff, whereas others may require catapults to reach flying speed for takeoff and then recover with a net or capture cable. Some small fixed-wing platforms such as AeroVironment's Raven™ are hand launched and recovered by stalling the aircraft over the intended land spot or by deploying a parachute.

2.2.2 Vertical Takeoff and Landing

A vertical takeoff and landing (VTOL) UAS platform has numerous applications. A VTOL platform can be in the form of a helicopter, a fixed-wing aircraft that can hover, or even a tilt-rotor. Some examples of a VTOL UAS would be the Northrop Grumman MQ-8 Fire Scout™ or the Bell Eagle Eye™ tilt-rotor (Figure 2.2). These UAS platforms have the advantage of small L&R footprints. This means that most do not need runways or roads to takeoff or land. Most also do not require any type of equipment such as catapults or nets for the L&R. Unlike fixed-wing platforms, the helicopter UAS can monitor from a fixed position requiring only a small space to operate.

Smaller electric helicopters, radio-control size, have advantages of very rapid deployment times making them ideal for search and rescue, disaster relief; or crime

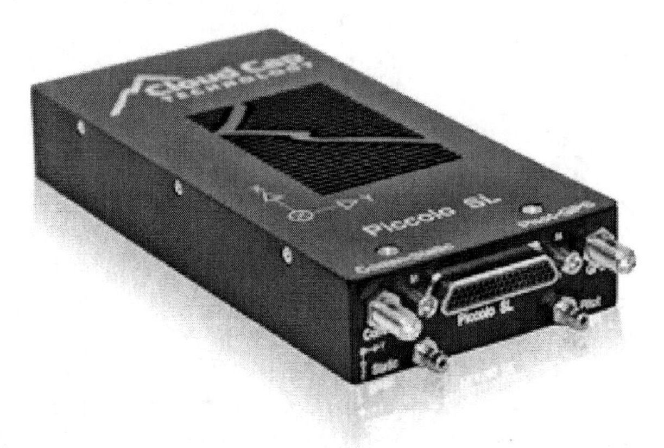

FIGURE 2.2 Piccolo™ SL autopilot unit. (Copyright Cloud Cap Technology, a Goodrich Company.)

fighting. Simple helicopter systems can be stored in a first responder's vehicle and launched within minutes. These small helicopters also offer the advantage of being somewhat covert when in operation at low altitudes. With no gasoline engine, the electric motor is quiet enough to allow it to operate at altitudes where it cannot be detected audibly. The disadvantages of small electric helicopters are that battery technology to date has not enabled long endurance to be achieved beyond 30 to 60 minutes.

2.3 COMMAND AND CONTROL ELEMENT

2.3.1 AUTOPILOT

The concept of autonomy is the ability for an unmanned system to execute its mission following a set of preprogrammed instructions without operator intervention. A fully autonomous UAS is able to fly without operator intervention from takeoff to touchdown. The amount of autonomy in a UAS varies widely from none to full autonomy. On one end of the spectrum the aircraft is operated completely by remote control with constant operator involvement (an external pilot). The aircraft's flight characteristics are stabilized by its autopilot system; however if the pilot were to be removed from the controls the aircraft would eventually crash.

On the other end of the spectrum the vehicle's onboard autopilot controls everything from takeoff to landing, requiring no pilot intervention. The pilot-in-command can intervene in case of emergencies, overriding the autopilot if necessary to change the flight path or to avoid a hazard. The autopilots for these vehicles are used to guide the vehicle along a designated path via predetermined waypoints.

Many commercial autopilot systems have become available in recent years for small UASs (sUASs). These small autopilot systems can be integrated to existing radio-controlled (hobby) aircraft or onto custom-built sUAS platforms. Commercial autopilot systems (often referred to as COTS for commercial-off-the-shelf systems; COTS is a widely used acronym for many different technologies) for sUAS have become smaller and lighter in recent years. They offer many of the same operational advantages that large RPA autopilots offer and are far less expensive. For example, the Cloud Cap Technology's Piccolo series of autopilots offers multivehicle control, fully autonomous takeoff and landing, VTOL and fixed-wing support, and waypoint navigation.

Autopilot systems for UASs are programmed with redundant technology. As a safety feature of most UAS autopilots, the system can perform a "lost-link" procedure if communication becomes severed between the ground control station and the air vehicle. There are many different ways that these systems execute this procedure. Most of these procedures involve creating a lost-link profile where the mission flight profiles (altitudes, flight path, and speeds) are loaded into the memory of the system prior to aircraft launch. Once the aircraft is launched, the autopilot will fly the mission profile as long as it remains in radio contact with the ground control station. The mission or lost-link profile can be modified when necessary if connectivity remains during flight. If contact with the ground station is lost in flight, the autopilot will execute its preprogrammed lost-link profile.

Other examples of lost link procedures include having the vehicle:

- Proceed to a waypoint where signal strength is certain in order to reacquire connectivity.
- Return to first waypoint and loiter or hover for a predetermined time in an attempt to reacquire the signal and then returning to landing waypoint to land if this is unsuccessful.
- Remain on current heading for a predetermined amount of time. During this time, any secondary means of communication can be attempted with the aircraft.
- Climb to reacquire link.
- Orbit where link was lost; at this point the external pilot then takes over using remote control technology, which uses VHF line-of-sight radio technology.

2.3.2 Ground Control Station

A ground control station or GCS is a land- or sea-based control center that provides the facilities for human control of unmanned vehicles in the air or in space (Figure 2.3). GCSs vary in physical size and can be as small as a handheld transmitter (Figure 2.4) or as large as a self-contained facility with multiple workstations. Larger military UASs require a GCS with multiple personnel to operating separate aircraft systems. One of the foremost goals for future UAS operation will be the capability for one crew to operate multiple aircraft from one GCS.

FIGURE 2.3 MQ-1 Predator GCS.

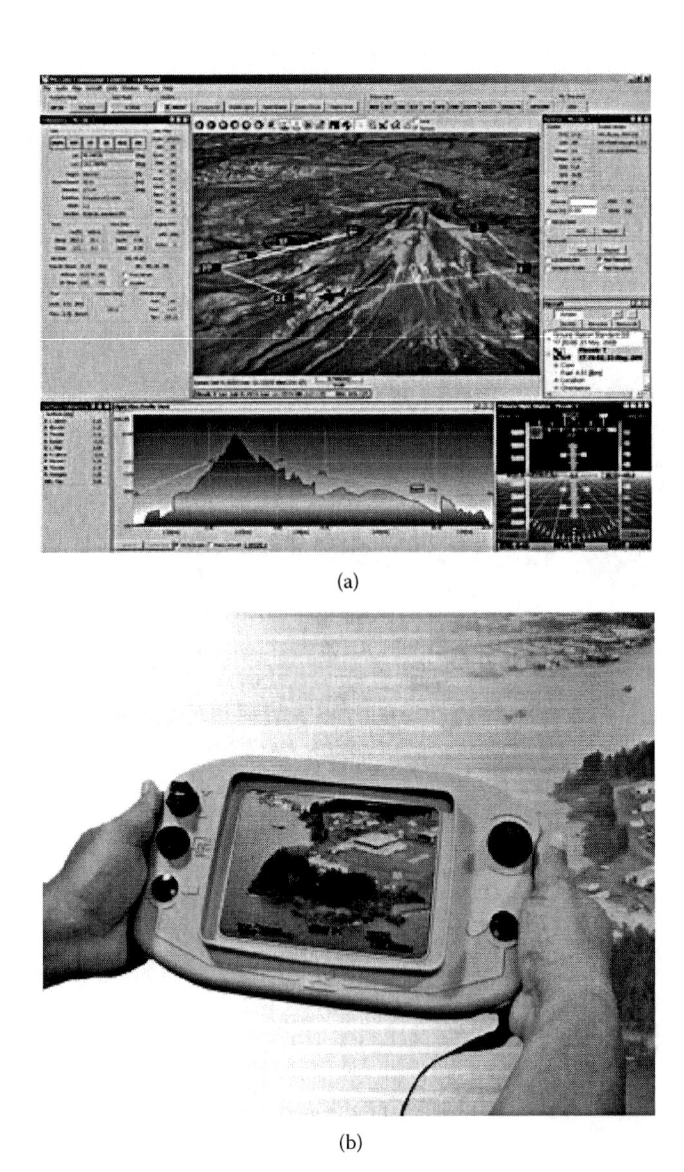

(a)

(b)

FIGURE 2.4 AeroVironment handheld GSC. (Copyright AeroVironment.)

A GSC usually consists of at least a pilot station and a sensor station. The pilot station is for just that: the pilot-in-command who operates the aircraft and its systems. The sensor station is for the operation of the sensor payload and radio communications. There can be many more operations, depending on the complexity of the UAS, which each require more workstations. For smaller less complex UASs these workstations may be combined requiring only one operator.

2.4 COMMUNICATION DATA LINK

Data link is the term used to describe how the UAS command and control information is sent and received both to and from the GCS and autopilot. UAS operations can be divided into two categories: radio frequency line-of-sight (LOS) and beyond line-of-sight (BLOS).

2.4.1 LINE-OF-SIGHT

Line-of-sight (LOS) operations refer to operating the RPA via direct radio waves. In the United States civilian LOS operations are usually conducted on the 915 MHz, 2.45 GHz, or the 5.8 GHz radio frequencies. These frequencies are unlicensed industrial, scientific, and medical (ISM) frequencies that are governed by Part 18 of the Federal Communications Commission (FCC) regulations. Other frequencies such as 310–390 MHz, 405–425 MHz, and 1350–1390 MHz are discrete LOS frequencies requiring a license on which to operate. Depending on the strength of the transmitter and receiver, and the obstacles in between, these communications can travel several miles. Signal strength can also be improved utilizing a directional tracking antenna. The directional antenna uses the location of the RPA to continuously adjust the direction in which it is pointed in order to always direct its signal at the RPA. Some larger systems have directional receiving antennas onboard the aircraft thereby improving signal strength even further.

ISM frequency bands are widely used making them susceptible to frequency congestion, which can cause the UAS to lose communication with the ground station due to signal interference. Rapid frequency hopping has emerged as a technology that minimizes this problem. Frequency hopping is a basic signal modulation technique used to spread the signal across the frequency spectrum. It is this repeated switching of frequencies during radio transmission that minimizes the effectiveness of unauthorized interception or jamming. With this technology, the transmitter operates in synchronization with a receiver, which remains tuned to the same frequency as the transmitter. During frequency hopping a short burst of data is transmitted on a narrowband, then the transmitter tunes to another frequency and transmits again, a process that repeats. The *hopping* pattern can be from several *times* per *second* to several thousand *times* per *second.* The FCC has allowed frequency hopping on the 2.45 GHz unlicensed band.

2.4.2 BEYOND LINE-OF-SIGHT

Beyond line-of-sight (BLOS) operations refer to operating the RPA via satellite communications or using a relay vehicle, usually another aircraft. Civilian operators have access to BLOS via the Iridium satellite system, which is owned and operated by Iridium LLC. Most sUASs do not have the need or ability to operate BLOS since their missions are conducted within line of sight range. Military BLOS operations are conducted via satellite on an encrypted Ku band in the 12 to 18 GHz range. One UAS in the market operates almost continuously through Ku band. Its launch phase is usually conducted using LOS and then transferred to BLOS data link. It is then transferred back to LOS for its recovery. One drawback of BLOS operations is that there can be several seconds of delay time once a command is sent to the aircraft, for it to respond to that command. This delay is caused by the many relays and systems it must travel through. With technological improvements over the past several years it is possible to conduct launch and recovery of the aircraft through BLOS data link.

2.5 PAYLOAD

Outside of research and development, most UASs are aloft to accomplish a mission and the mission usually requires an onboard payload. The payload can be related to surveillance, weapons delivery, communications, aerial sensing, or cargo. UASs are often designed around the intended payload they will employ. As we have discussed, some UASs have multiple payloads. The size and weight of payloads is one of the largest considerations when designing a UAS. Most commercial application sUAS platforms require a payload less than 5 lbs. Manufactures of some sUAS have elected to accommodate interchangeable payloads that can be quickly removed and replaced.

In reference to the missions of surveillance and aerial sensing, sensor payloads come in many different forms for different missions. Examples of sensors can include electro-optical (EO) cameras, infrared (IR) cameras, synthetic aperture radars (SAR), or laser range finder/designators. Optical sensor packages (cameras) can be either installed by permanently mounting them to the UAS aircraft giving the sensor operator a fixed view only, or they can employ a mounted system called a gimbal or turret (Figure 2.5). A gimbal or turret mounting system gives the sensor a predetermined range of motion usually in two axes (vertical and horizontal). The gimbal or turret receives input either through the autopilot system or through a separate receiver. Some gimbals are also equipped with vibration isolation, which reduces the amount of aircraft vibration that is transmitted to the camera thus requiring less electronic image stabilization to produce a clear image or video. Vibration isolation can be performed by either an elastic/rubber mounting or using an electronic gyrostabilization system.

2.5.1 Electro-Optical

Electro-optical cameras are so named because they use electronics to pivot, zoom, and focus the image. These cameras operate in the visible light spectrum. The imagery they yield can be in the form of full motion video, still pictures, or even blended still and video images. Most sUAS payload EO cameras use narrow to mid field of view (FOV) lenses. Larger UAS camera payloads can also be equipped with wide or ultrawide FOV (WFOV) sensors. An EO sensor can be used for many missions and combined with different types of sensors to create blended images. They are most frequently operated during daylight hours for optimal video quality.

2.5.2 Infrared

Infrared cameras operate in the infrared range of the electromagnetic spectrum (approximately 1–400 THz). IR, or sometimes called FLIR for forward-looking

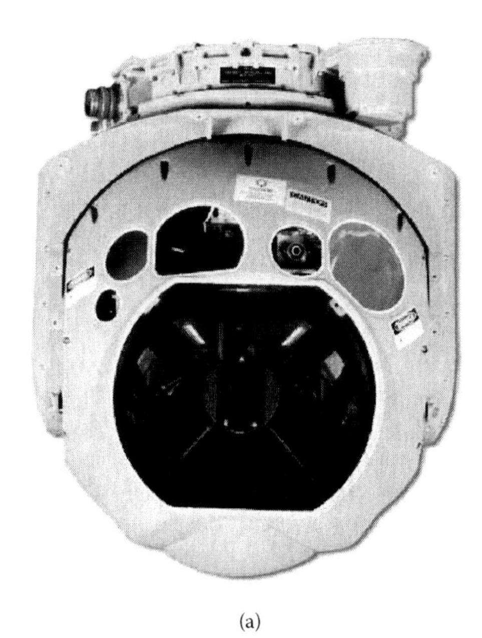

(a)

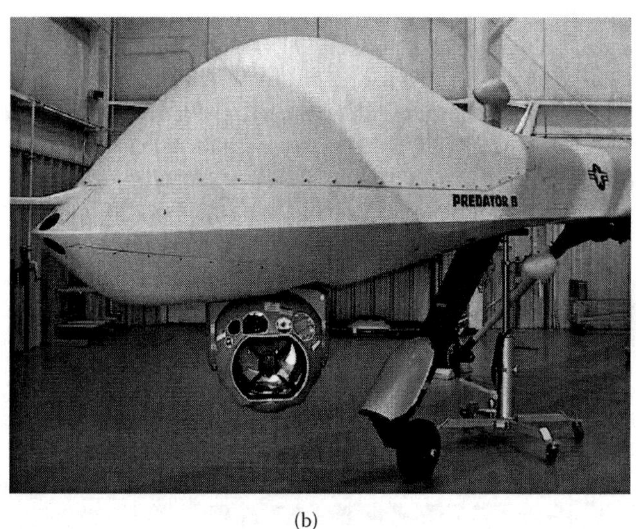

(b)

FIGURE 2.5 (a) Cloud Cap Technology Tase200 EO/IR gimbal mounted on a (b) General Atomics MQ-1. ([a] Copyright Cloud Cap Technology. [b] Copyright General Atomics.)

infrared, sensors form an image using IR or heat radiation. Two types of IR cameras used for UAS payloads are cooled and noncooled. Cooled cameras are often more expensive and heavier than noncooled cameras. Modern cooled cameras are cooled by a cryocooler which lowers the sensor temperature to cryogenic temperature (below 150°C). These systems can be manufactured to produce images in the midwave infrared (MWIR) band of the spectrum where the thermal contrast is high. These types of cameras can also be designed to work in the longwave infrared (LWIR) band. A cooled IR camera's detectors are typically located in a vacuum sealed case and require extra power to cool. In general, cooled cameras produce a higher quality image than uncooled cameras.

Noncooled cameras use sensors that are at or just below ambient temperature and work through the change of resistance, voltage, or current created when heated by the infrared radiation it detects. Noncooled sensors are designed to work in the LWIR band from 7 to 14 microns in wavelength, where terrestrial temperature targets emit most of their infrared energy.

2.5.3 Laser

A laser range finder uses a laser beam to determine the distance to an object. A laser designator uses a laser beam to designate a target. The laser designator sends a series of invisible coded pulses that reflect back from the target and are detected by the receiver. There are, however, drawbacks to using a laser designator on an intended target. The laser may not be accurate if atmospheric conditions are not clear, such as rain, clouds, blowing dust, or smoke. The laser can also be absorbed by special paints or reflect incorrectly or not at all such as when aimed at glass.

2.6 LAUNCH AND RECOVERY

The launch and recovery element (LRE) of the UAS is often one of the most labor-intensive aspects of the UAS operation. Some UASs have very elaborate LRE procedures, whereas others have virtually none. Larger systems have procedures and dedicated personnel that prepare, launch, and recover the RPA. Runway lengths of up to 10,000 feet and support equipment such as ground tugs, fuel trucks, and ground

FIGURE 2.6 Soldier hand-launching a Raven RQ-11. (Copyright AeroVironment.)

(a)

(b)

FIGURE 2.7 An unmanned aerial vehicle ScanEagle lands in the skyhook for recovery on the flight deck aboard the amphibious assault ship *USS Saipan*.

power units or GPUs are needed for these large UASs. Small VTOL UASs tend to have the least complex procedures and equipment when it comes to LRE, most of which consists of only a suitable takeoff and landing area. Other UASs, such as the Raven manufactured by AeroVironment Inc., have very small LRE since they can be hand launched and recovered by an onboard parachute (Figure 2.6).

There are many ways to conduct launch and recovery operations for current UASs. Some of the most common involves using a catapult system to get the aircraft to flight speed in a very short distance. The ScanEagle™ manufactured by Isitu, a Boeing company, utilizes a catapult for takeoff and an arresting cable Isitu calls the SkyHook™ for its recovery (Figure 2.7). In this system the vehicle is equipped with a hook on the end of its wing tips as well as employing a very precise dual global positioning system to fly into the suspended cable for recovery.

The Aerosonde Mark 4.7 manufactured by Aerosonde has optional LRE equipment. It can be launched using a car top launcher whereby a ground vehicle is used

FIGURE 2.8 AAI Aerosonde vehicle top-launch cradle. (Copyriht AAI Aerosonde.).

to enable the UAS to reach flying speed (Figure 2.8). A catapult system is also available for the Aerosonde. For the landing phase it can "belly land" on grass or hard surfaces, or it can recover into a moving net.

2.7 HUMAN ELEMENT

The most important element of the UAS is the human element. At this point the human element is required for the operation of the UAS. This element consists of a pilot, a sensor, and supporting ground crew. As previously mentioned, some of these positions can be combined into one depending on the complexity of the system. In the future, the human element will likely get smaller as technological capability increases. As with commercial airliners of the past, automation will require less human interaction. The UAS pilot in command is responsible for the safe operation of the aircraft. This element is described in greater detail in Chapter 11.

3 U.S. Aviation Regulatory System

Douglas M. Marshall

CONTENTS

3.1 U.S. AVIATION REGULATORY SYSTEM

3.1.1 INTRODUCTION

Aviation regulations in the United States have existed nearly as long as the technology that is being regulated. All levels of government in civilized countries impose various regulations on their citizens and their activities.

Regulations in any technical environment such as aviation are typically driven by original equipment manufacturers (OEMs) and operators. As users experience incidents, problems, or anomalies, those events are properly reported to the Federal Aviation Administration (FAA). Should the number of events reach a certain critical mass or the outcome is sufficiently severe (fatalities, injuries, or property damage), the data generated may provoke a review of the relevant regulation, if any.

The introduction of a new technology or procedure into the National Airspace System (NAS) requires a comprehensive safety analysis before the FAA can allow the change. The safety analysis includes a review of the relevant regulation and supporting advisory circulars (AC) or special Federal Aviation Regulations (SFARs) to determine whether the proposed technology or procedure can comply with current regulation. The FAA may grant exceptions in particular cases after performing the safety review as a way to manage unique, perhaps nonrecurring circumstances, or when the event that led to the review is determined to be unlikely to recur.

These circumstances may lead to the rulemaking process, which provides the mechanism for the FAA to fulfill its statutory mandate to ensure the safety of the aviation environment. This chapter describes the history of the U.S. federal as well as international aviation regulations; the structure of those statutes and regulations; the purpose and intent of the rules; how rules are made, changed and enforced; how this system affects the development of unmanned aircraft system (UAS) technologies and operations; and concludes with a look into the future of UAS regulations.

3.1.2 History of U.S. Aviation Regulation

Aviation regulations in the United States have enjoyed a long and colorful history, beginning with the commencement of airmail operations by the U.S. Post Office in 1918, only 15 years after the first manned powered flight. Three years before that President Wilson signed a bill that created a National Advisory Committee on Aeronautics, which was intended to oversee a scientific study of the "problems" of flight. No fewer than six federal statutes enacted to regulate some aspect of aviation followed these early efforts. Most were directed toward safety concerns and the perceived need to bring some order to the commercial aspects of aviation. The issues that generated the greatest concern were the number of crashes, the need for a regulated civil airport network, the lack of a harmonized or common system of air navigation, and demand for a civil aviation infrastructure that would support growth and stability of the industry, both for military and nonmilitary applications.

3.1.3 Federal Aviation Administration

The Federal Aviation Administration was created by the Federal Aviation Act of 1958.* The statute was enacted in response to a series of fatal accidents and midair collisions involving commercial passenger aircraft. The FAA is part of the Department of Transportation, and derives its rulemaking and regulatory power from Title 49 of the United States Code, Section 106. The Commerce Clause of the U.S. Constitution (Article I, Section 8) grants Congress broad authority to "regulate commerce with foreign nations, and among the several states." The U.S. government therefore has exclusive power to regulate the airspace of the United States.† A citizen of the United States has a public right of transit through the navigable airspace.‡ Among other powers the statute confers upon the administrator of the FAA is the mandate to develop plans and policy for the use of the navigable

* Public Law 85-726, 85th Cong., 2nd Sess.; 72 Stat. 731; 49 U.S.C § 1301, as amended.
† 49 U.S.C. § 40103 (a)(1).
‡ 49 U.S.C. § 40103 (a)(2).

airspace and assign by regulation or order the use of the airspace necessary to ensure the safety of aircraft and the efficient use of airspace.[*] The administrator may modify or revoke a regulation, order, or guidance document when required in the public interest. The administrator shall prescribe air traffic regulations on the flight of aircraft (including regulations on safe altitudes) for navigating, protecting, and identifying aircraft; protecting individuals and property on the ground; using the navigable airspace efficiently; and preventing collision between aircraft, between aircraft and land or water vehicles, and between aircraft and airborne objects.[†]

Pursuant to its rulemaking authority, the FAA has set forth the standards for the operation of aircraft in the sovereign airspace of the United States.[‡] Commonly known as the FARs (Federal Aviation Regulations), these regulations are the "rules of the road" for certification of all civil aircraft,[§] airmen,[¶] and airspace[**]; certification and operations for air carriers and operators for compensation or hire[††]; air traffic and general operating rules[‡‡]; and schools and other certificated agencies,[§§] airports,[¶¶] and navigational facilities.[***]

The first section of 14 CFR, Part 1.1, lists the definitions and abbreviations to be observed in the ensuing parts and subparts of the FARs. Of more than passing interest to the unmanned aircraft community is the fact that the terms *UAV* or *UAS* or *unmanned system* or *unmanned* aircraft or any other term referring to remotely piloted aircraft are from the FARs or, for that matter, any other federal regulation or statute. The term *aircraft* is defined as "a device that is used or intended to be used for flight in the air."[†††] Similarly, "*airplane* means an engine-driven fixed-wing aircraft heavier than air, that is supported in flight by the dynamic reaction of the air against its wings."[‡‡‡] "*Air traffic* means aircraft operating in the air or on an airport surface, exclusive of loading ramps and parking areas."[§§§]

The FAA regulates aircraft, airmen, certain categories of employees of airlines and commercial or common carrier operations, airports, and the national airspace. The FAA's "toolbox" is the system of regulations, rulemaking processes, certifications, advisory circulars, special authorizations, and directives that the agency uses to carry out its regulatory functions of rulemaking, surveillance, compliance, and enforcement.

Three of the tools that the FAA uses to administer the FARs are advisory circulars, airworthiness directives, and policy statements. An advisory circular (AC) or

* 49 U.S.C. § 40103 (b)(1).
† 49 U.S.C. § 40103 (b)(2).
‡ 14 CFR Part 1.1 et seq.
§ 14 CFR Parts 21–49.
¶ 14 CFR Parts 61–67.
** 14 CFR Parts 71–77.
†† 14 CFR Parts 119–135.
‡‡ 14 CFR Parts 91–105.
§§ 14 CFR Parts 141–147.
¶¶ 14 CFR Parts 150–161.
*** 14 CFR Parts 170–171.
††† 14 CFR 1.1.
‡‡‡ 14 CFR 1.1.
§§§ 14 CFR 1.1.

airworthiness directive (AD) may be issued in response to a safety-related event or system anomaly, or a technical standards order (TSO) could be developed to remediate a technical problem. An AC provides guidance to owners or operators of aircraft or systems to facilitate compliance with the applicable regulations. An AD is a notification to owners and operators of certified aircraft that a known safety deficiency with a particular model of aircraft, engine, avionics, or other system exists and must be corrected. A TSO is a minimum performance standard for specified materials, parts, and appliances used on civil aircraft. When authorized to manufacture a material, part, or appliance to a TSO standard, this is referred to as TSO authorization. Issuance of a TSO authorization constitutes both design and production approval. However, issuance of a TSO authorization is not an approval to install and use the article in the aircraft. It simply means that the article meets the specific TSO and the applicant is authorized to manufacture it.

Advisory circulars are utilized to advise the aviation community on issues pertaining to the regulations, but are not binding upon the public. The exception would be when an advisory circular is specifically referenced in a regulation.* The advisory circulars are issued in a numbered-subject system corresponding to the subject areas of the FARs.† The advisory circular that has created the most controversy in the unmanned aviation world is AC 91-57, which will be discussed in more detail later. That circular references 14 CFR Part 91 (Air Traffic and General Operating Rules), which contains the airspace regulations.

Another advisory tool is the policy statement. Administrative implementation (as announced or documented by a published policy statement) of a particular statutory provision shall be accorded deference by the courts when it appears that Congress delegated authority to the agency generally to make rules carrying the force of law, and that the agency interpretation claiming deference was promulgated in the exercise of that authority. Delegation of such authority may be shown in a variety of ways, as by an agency's power to engage in adjudication or notice-and-comment rulemaking, or by some other indication of a comparable congressional intent.‡ The FAA has issued three policy statements pertaining to unmanned aircraft, AFS-400 UAS Policy Statement 05-01; a clarification published in the *Federal Register* February 6, 2007, titled "Unmanned Aircraft Operations in the National Airspace System"; and "Interim Operational Approval Guidance 08-01," which likewise references 14 CFR Part 91.§

3.1.4 Enforcement and Sanctions

No system of rules regulations can be effective without a means to enforce them. FARs are no exception. The FAA's mandate from Congress is to conduct surveillance of aviation activities, inspect aviation systems, investigate violations of the aviation regulations, and take appropriate measures to enforce the regulations in the event of

* Advisory Circular 00-2.11 (1997).
† Advisory Circular 00-2.11 (1997).
‡ *United States v. Mead Corp.*, 533 U.S. 218; 121 S. Ct. 2164; 150 L. Ed. 2d 292.
§ 72 FR 6689, Vol. 72, No. 29, February 13, 2007 (Interim Operational Approval Guidance 08-01).

a violation. The agency's investigative power extends to all provisions of the Federal Aviation Act of 1958; the Hazardous Materials Transportation Act; the Airport and Airway Development Act of 1970; the Airport and Airway Improvement Act of 1982; the Airport and Airway Safety and Capacity Expansion Act of 1987; and any rule, order, or regulation issued by the FAA. The FAA's central mission, pursuant to its own Order 2150.3A (Compliance and Enforcement Program) is to promote adherence to safety standards, but the agency recognizes that, due to the nature of aviation itself, it must largely depend upon voluntary compliance with the regulatory standards. The Fifth and Fourteenth Amendments to the U.S. Constitution require that the FAA enforcement process provide "due process" in the procedures for ensuring compliance with the regulations. This means that no one shall be deprived of "life, liberty, or property without due process of law."* Thus, the FAA may not act arbitrarily or inconsistently in its efforts to enforce the regulations.

The enforcement process established by the FAA is designed to be fair, reasonable, and perceived to be fair by those who are subject to the regulations. It is a complicated process that provides a number of decision points that allow the FAA and the party being investigated to arrive at an informal resolution rather than taking the matter to a fully litigated trial. The range of possible outcomes varies from a case being abandoned by the FAA after it investigates an alleged violation to a trial, an outcome, and an appeal to the U.S. Court of Appeals or even to the U.S. Supreme Court (a rare event indeed). Trials are conducted like any other civil trial, and the FAA generally has the burden of proof in establishing a violation. Short of a civil penalty or certificate revocation or suspension, the FAA may issue warning letters or letters of correction, which are intended to bring the alleged offender into compliance with the regulations for violations that are not deemed sufficiently serious to warrant more severe sanctions. A constructive attitude of cooperation by the certificate holder often goes a long way toward resolving inadvertent or nonflagrant violations by first-time offenders. Civil penalties of up to $50,000 per violation are available in cases where there is no certificate to suspend or revoke, where revocation would impose an undue hardship, where qualification is not at issue, or where the violation is too serious to be handled administratively by use of remedial action. It is important to remember that the lack of an airman's certificate or other FAA-issued license does not immunize a person or entity from an FAA enforcement action backed up by imposition of a civil penalty.

The manner in which the FAA has chosen to enforce the FARs when dealing with unmanned aircraft will be dealt with below. As of this publication, the authors are not aware of any formal enforcement action that the FAA has taken against any unmanned aircraft system/remotely piloted aircraft operator, pilot, owner, manufacturer, or service.

3.2 INTERNATIONAL AVIATION REGULATIONS

As early as 1919, an international agreement (the Convention for the Regulation of Aerial Navigation, created by the Aeronautical Commission of the Peace Conference

* U.S. Constitution, Amendment 5 and Amendment 14.

of 1919, otherwise known as the Versailles Treaty) recognized that the air above the high seas was not as "free" as the water of those seas. In that convention the contracting states acknowledged exclusive jurisdiction in the airspace above the land territory and territorial waters of the states, but agreed to allow, in times of peace, innocent passage of the civil aircraft of other states, so long as the other provisions of the convention were observed. States still retained the right to create prohibited areas in the interests of military needs or national security. During the global hostilities of the 1940s the United States initiated studies and later consulted with its major allies regarding further harmonization of the rules of international airspace, building upon the 1919 convention. The U.S. government eventually extended an invitation to 55 states or authorities to attend a meeting to discuss these issues, and in November 1944, an International Civil Aviation Conference was held in Chicago. Fifty-four states attended this conference, at the end of which the Convention on International Civil Aviation was signed by 52 of those states. The convention created the permanent International Civil Aviation Organization (ICAO) as a means to secure international cooperation and the highest possible degree of uniformity in regulations and standards, procedures, and organization regarding civil aviation matters. The Chicago Conference laid the foundation for a set of rules and regulations regarding air navigation as a whole, which was intended to enhance safety in flying and construct the groundwork for the application of a common air navigation system throughout the world.

The constitution of the ICAO is the Convention on International Civil Aviation that was drawn up by the Chicago conference, and to which each ICAO contracting state is a party. According to the terms of the convention, the organization is made up of an assembly, a council of limited membership with various subordinate bodies, and a secretariat. The chief officers are the president of the council and the secretary general. ICAO works in close cooperation with other members of the United Nations family such as the World Meteorological Organization, the International Telecommunication Union, the Universal Postal Union, the World Health Organization, and the International Maritime Organization. Nongovernmental organizations that also participate in ICAO's work include the International Air Transport Association, the Airports Council International, the International Federation of Airline Pilots' Associations, and the International Council of Aircraft Owner and Pilot Associations.*

ICAO's objectives are many and are set forth in the 96 articles of the Chicago Convention and the 18 annexes thereto. Additional standards and guidelines are found in numerous supplements (Standards and Recommended Practices, or SARPS) and Procedures for Air Navigation Services, which are under continuing review and revision. Contracting states are free to take exceptions to any element of the annexes, and those exceptions are also published. Contracting States are also responsible for developing their own aeronautical information publication (AIP), which provide information to ICAO and other states about air traffic, airspace, airports, navaids (navigational aids), special use airspace, weather and other relevant data for use by air crews transiting into or through the state's airspace. The AIPs will also contain

* International Civil Aviation Organization: http://www.icao.int/.

information about a state's exceptions to the annexes and any significant differences between the rules and regulations of the state and ICAO's set of rules.

The annexes cover rules of the air, meteorological services for international air navigation, aeronautical charts, measurement units used in air and ground operations, operation of aircraft, aircraft nationality and registration marks, airworthiness of aircraft, facilitation (of border crossing), aeronautical communications, air traffic services, search and rescue, aircraft accident investigation, aerodromes, aeronautical information services, environmental protection, security-safeguarding international civil aviation against acts of unlawful interference, and the safe transportation of dangerous goods by air. The only reference in any ICAO document to unmanned aircraft is found in Article 8 of the convention, which states that:

> No aircraft capable of being flown without a pilot shall be flown without a pilot over the territory of a contracting State without special authorization by that State and in accordance with the terms of such authorization. Each contracting State undertakes to insure that the flight of such aircraft without a pilot in regions open to civil aircraft shall be so controlled as to obviate danger to civil aircraft.

ICAO's rules apply to international airspace, which is typically defined as the airspace over the high seas more than 12 miles from the sovereign territory of a state (country) as well as some domestic airspace by virtue of incorporation into a contracting state's own regulatory scheme. The rules apply to all contracting states (there are 188 of them), so any nation that elects not to become an ICAO member is not entitled to the protection of ICAO's rules. However, ICAO is a voluntary organization, and there are no provisions for enforcement of the regulations or standards such as those found in the FARs. As a founding member of ICAO and as a nation that has a substantial interest in preserving harmony in international commercial aviation, the United States enforces ICAO's rules against U.S. operators to the extent that the ICAO rule has been incorporated into the FARs and does not conflict with domestic regulations.

Additional international aviation organizations located in Europe that exercise some level of regulatory powers include EUROCONTROL (European Organization for the Safety of Air Navigation), EASA (European Aviation Safety Agency), and EUROCAE (European Organization for Civil Aviation Equipment). EUROCONTROL is an intergovernmental organization that acts as the core element of air traffic control services across Europe, and is dedicated to harmonizing and integrating air navigations services in Europe and creating a uniform air traffic management system for civil and military users. The agency accomplishes this by coordinating efforts of air traffic controllers and air navigation providers to improve overall performance and safety. The organization is headquartered in Brussels and has 38 member states. The European Commission created a Single European Sky ATM Research (SESAR) initiative in 2001 and has delegated portions of the underlying regulatory responsibility to EUROCONTROL. EASA was established as an agency of the European Union in 2003 and has regulatory responsibility in the realm of civilian aviation safety, assuming the functions formerly performed by the Joint Aviation Authorities (JAA). Contrary to JAA's role, EASA has legal regulatory authority, which includes enforcement power. EASA has responsibility for airworthiness and environmental certification of aeronautical products manufactured, maintained or used by persons under the regulatory oversight of European Union member states. EUROCAE reports to EASA, although it was created many years before EASA was formed, and deals exclusively with aviation standardization (with reference to airborne and ground systems and equipment). Its membership is made up of equipment and airframe manufacturers, regulators, European and international civil aviation authorities, air navigation service providers, airlines, airports, and other users. EUROCAE's Working Group 73 is devoted to the development of products intended to help assure the safe, efficient, and compatible operation of UASs with other vehicles operating within nonsegregated airspace. WG-73 makes recommendations to EUROCAE with the expectation that those recommendations will be passed on to EASA.

In addition to ICAO and the three European organizations just discussed, any nation's civil aviation authority (CAA) is free to promulgate its own aviation rules and regulations for operations within their sovereign airspace, and until ICAO has created an overarching set of rules for UAS operations among its member states, operators of UAS must be sensitive to the rules and regulations of the contracting state that is providing air traffic services in international airspace.

3.3 STANDARDS VERSUS REGULATIONS

The FAA exercises its statutory mandate by making rules and regulations. Those efforts are often supplemented or enhanced by published standards that are created by industry organizations and approved by the FAA. Standards developers work with engineers, scientists, and other industry personnel to develop nonbiased standards or specification documents that serve industry and protect the public. These developers can be private concerns, trade organizations, or professional societies. Standards providers are distributors of codes, standards, and regulations. They may also provide access to a database of standards. The supplier may or may not be the developer of the standards distributed.

These organizations are essentially professional societies made up of industry representatives, engineers, and subject matter experts who provide advisory support to federal agencies such as the FAA. They make recommendations that may become a formal rule by adoption or reference. Engineering codes, standards, and regulations all serve to ensure the quality and safety of equipment, processes, and materials. The three most prominent of those advisory organizations playing a role in the evolution of unmanned aviation are the Society of Automotive Engineers (SAE), the Radio Technical Commission for Aeronautics (RTCA), and ASTM International (originally the American Society of Testing and Materials).

Aeronautical engineering codes are enforced by the FAA and are critical to developing industry practices. Whereas engineering regulations such as those found in the FARs are government-defined practices designed to ensure the protection of the public as well as uphold certain ethical standards for professional engineers, engineering standards ensure that organizations and companies adhere to accepted professional practices, including construction techniques, maintenance of equipment, personnel safety, and documentation. These codes, standards and regulations also address issues regarding certification, personnel qualifications, and enforcement.

Manufacturing codes, standards, and regulations are generally designed to ensure the quality and safety of manufacturing processes and equipment, and aviation regulations are no exception. Manufacturing standards ensure that the equipment and processes used by manufacturers and factories are safe, reliable, and efficient. These standards are often voluntary guidelines but can become mandatory by reference in the FARs. Manufacturing regulations are government-defined and usually involve legislation for controlling the practices of manufacturers that affect the environment, public health, or safety of workers. Aircraft manufacturers in the United States and European Union are required by law to produce aircraft that meet certain airworthiness and environmental emissions standards.

The FAA has supported and sponsored four domestic committees dedicated to developing standards and regulations for the manufacture and operation of unmanned aircraft. RTCA's Special Committee 203 Unmanned Aircraft Systems (SC-203) began developing minimum operational performance standards (MOPS) and minimum aviation system performance standards (MASPS) for unmanned aircraft systems in 2004: "SC-203 products will help assure the safe, efficient and compatible operation of UAS with other vehicles operating within the NAS. SC-203 recommendations will be based on the premise that UAS and their operations will not have a negative impact on existing NAS users."

ASTM's F-38 Unmanned Air Vehicle Systems Committee addresses issues related to design, performance, quality acceptance tests, and safety monitoring for unmanned air vehicle systems. Stakeholders include manufactures of unmanned aerial vehicles and their components, federal agencies, design professionals, professional societies, maintenance professionals, trade associations, financial organizations, and academia.

SAE's G-10U Unmanned Aircraft Aerospace Behavioral Engineering Technology Committee was established to generate pilot training recommendations for unmanned aircraft systems civil operations and has released the recommendations.

By Order 1110.150 signed on April 10, 2008, the FAA created a Small Unmanned Aircraft System (sUAS) Aviation Rulemaking Committee (ARC) according to the FAA administrator's authority under Title 49 United States Code (49 U.S.C.) § 106(p)(5). The committee's term was 20 months and was made up of representatives of aviation associations, industry operators, manufacturers, employee groups or unions, the FAA and other government entities, and other aviation industry participants, including academia. The committee delivered its formal recommendations to the FAA associate administrator in March 2009. The FAA's Air Traffic Organization simultaneously convened a safety risk management (SRM) committee that was charged with describing the UAS systems under review, identifying hazards, analyzing risk, assessing risk, and treating risk to arrive at a safety management system (SMS) for UAS that would be coordinated or integrated with the ARC's recommendations. This process follows a number of FAA policies that require oversight and regulation of aeronautical systems that may impact safety in the National Airspace System (FAA Order 8000.369 Safety Management System Guidance; FAA Order 1100.161 Air Traffic Safety Oversight; FAA Order 8000.36 Air Traffic Safety Compliance

Process; FAA Order 1000.37 Air Traffic Organization Safety Management System Order; ATO-SMS implementation Plan Version 1.0, 2007; FAA SMS Manual Version 2.1 of June 2008; Safety and Standards Guidance Letter 08-1; and AC 150/5200-37 Introduction to SMS for Airport Operations).

The ARC's recommendations for regulations pertaining to small UAS are under review as of this publication, but will eventually lead to a published notice of proposed rulemaking (discussed later). This will be the first set of regulations to be proposed by the FAA dealing specifically with unmanned aircraft systems.

One previous attempt to address a narrow category of remotely piloted aircraft was Advisory Circular 91-57, published in 1981. This AC was in reality an effort to regulate by not regulating the recreational modeling community, outlining and encouraging voluntary compliance with safety standards for model aircraft operators. The document's content was taken off the FAA's website, but it has not been revoked, so it remains as the operative standard for model aircraft operations within certain designated areas and under the authority of a voluntary organization, the Academy of Model Aeronautics (AMA). The AMA created its own set of standards and restrictions for its members, compliance with which is a prerequisite for the group insurance coverage for which its members are eligible.

Although AC 91-57 was specifically directed toward recreational modelers, the circular has at times been relied upon by commercial UAS operators and developers to make a claim that they can fly their small UAS under 400 feet AGL without communicating with the FAA and running afoul of the FARs. Policy Statement 05-01 and Guidance Document 08-01 both refer to AC 91-57 as the official policy with respect to recreational and hobbyist aero modeling, which is that those activities do not fall under the intent of FARs and are thus excluded. However, by inference, the FAA believes that it has the statutory power to regulate recreational models because they fall under the definition of *airplane* found in 14 CFR 1.1, but chooses not to do so as a matter of policy.

3.4 HOW THE PROCESS WORKS

The sUAS ARC discussed earlier is an example of one of the FAA's processes for creating the rules, regulations, circulars, directives, and orders that it employs to bring some order to the aviation industry, which is one of the most heavily regulated industries in the United States and elsewhere. The FAA's rulemaking authority is derived from either executive order (from the Office of the President) or the U.S. Congress, through specific mandate or by delegation of Congress' lawmaking powers as conferred by the U.S. Constitution (Article 1, Section 8). The FAA relies on those two sources as well as recommendations from the National Transportation Safety Board, the public, and the FAA itself to initiate rulemaking. Ultimately the FAA makes rules to serve the public interest and to fulfill its mission of enhancing safety in the aviation environment.

The process of rulemaking is governed by the Administrative Procedures Act of 1946 and the Federal Register Act of 1935. These two statutes combined were intended to ensure that the process is open to public scrutiny (that federal agencies do not make rules or impose regulations in secret or without full transparency).

This is accomplished by procedural due process and publication requirements. This "informal rulemaking" is a four step process that follows what often involves months or even years of industry rulemaking committee effort, internal FAA review and analysis, and interagency negotiation. Once the proposed rule has achieved a sufficient level of maturity, it will be published in the *Federal Register* as a "Notice of Proposed Rulemaking." The notice provides an opportunity to the general public to comment on the proposed rule within a certain period of time. The comments from the public must be resolved in some fashion before the final rule document is published, which should respond to the comments and provide an explanation of purpose and basis for rule as well as the way in which the comments were resolved. The last step is implementation, and the effective date must be at least 30 days after publication of the final rule unless it is interpretive, a direct rule, a general policy statement, an emergency rule, or a substantive rule that grants an exemption to an existing rule or requirement. Some agency rules or policies may be exempted from this process, such as interpretive rules or general policy statements, or if the agency can demonstrate that the notice and comment process would be impractical, unnecessary, or contrary to public interest (showing "good cause").

Rules that have gone through this informal rulemaking process have the same force and effect as a rule or regulation imposed by an act of Congress. Thus, the FAA is empowered to enforce those rules as if they were laws enacted by Congress. The rules are typically referenced to or codified in the Code of Federal Regulations (CFRs). There are exceptions to this, however. Direct final rules are implemented after a final rule is issued while still providing for a period of notice and comment. The rule becomes effective after the specified period if there are no adverse comments. The difference in this process from the notice of proposed rulemaking procedure is that there is no proposed rule published before the final rule is released. This is used for routine rules or regulations that are not anticipated to generate comment or controversy. Interim rules are usually effective immediately and are issued without prior notice, often in response to an emergency. A final rule may issue based upon the interim rule after a period of comment. The status of an interim rule as final or amended or withdrawn is always published in the *Federal Register*. Last, interpretive rules may be issued to explain current regulations or its interpretation of existing statues or rules. This tool is not commonly used by the FAA, but may be useful when a rule is repeatedly misinterpreted, resulting in chronic compliance issues.

The point of this complex, sometimes cumbersome, and time-consuming process of rulemaking is to advance the cause of safety and harmonization so that all users and others affected by the aviation environment are protected from undue risk of harm. Further it is to ensure that all entities operate under the same set of rules and regulations, and have abundant opportunity to engage in the process so that the outcomes may be influenced by multiple points of view. Each step of the process requires a series of reviews by other agencies of the federal government, such as the Office of the Secretary of Transportation, the Department of Transportation, the Office of Management and Budget, the General Accountability Office and the Office of the Federal Register. A flow chart of the aviation regulatory process would demonstrate at least a 12-step process, with multiple interim steps imbedded in most of the broader categories. If a proposed rule were to be subjected to each and every possible review step, the list would include no fewer than 35 stops along the way. For example, due to the many and equally influential stakeholders that could be involved in an effort by the Department of Defense to create a new restricted area for UAS operations, testing, and training, it is commonly estimated that it would take 5 years to accomplish the goal. As an example, the implementation of an aviation safety device for air transport aircraft known as TCAS (traffic alert and collision avoidance system) required over 15 years from inception to implementation, and it took an act of Congress to mandate the use of TCAS in commercial airliners.

In addition to formal rules and regulations, the FAA issues orders, policies, directives, and guidance documents. The FAA routinely issues policy statements and guidance documents to clarify or explain how the FAA interprets and enforces the regulations. A policy statement gives guidance or acceptable practices on how to find compliance with a specific CFR section or paragraph. These documents are explanatory and not mandated. They are also not project specific. Practically speaking, this means that they are not enforceable in formal compliance proceedings, but they do provide guidance to users and the public on how best to comply with the FARs. Guidance documents are similar in nature, and are likewise explanatory rather than mandatory.

The FAA's website contains links to all historical and current policy statements, guidance documents, orders, directives, circulars and regulations. Binding orders and regulations are published in the *Federal Register* and are accessible on the Electronic Code of Federal Regulations (e-CFR) government Web site.*

3.5 CURRENT REGULATION OF UNMANNED AIRCRAFT

As discussed earlier, there is no specific reference in any of the Federal Aviation Regulations to unmanned aircraft, pilots/operators of unmanned aircraft, or operations in the national airspace of unmanned aircraft. A literal reading of the definitions listed in 14 CFR 1.1 would include all unmanned aircraft in the description of *aircraft*. There is no case authority, nor is there a rule or regulation that says that unmanned aircraft of any size or capability are *not* regulated. This conceivably would include radio-controlled model aircraft. In recognition of the reality that radio-controlled aircraft are aircraft but not of the type that the FAA is inclined to regulate, Advisory Circular 91-57 was published in 1981. This AC encourages voluntary compliance with safety standards for model aircraft operators. The circular also acknowledges that model aircraft may pose a safety hazard to full-scale aircraft in flight and to persons and property on the ground.† Modelers are encouraged to select sites that are sufficiently far away from populated areas so as to not endanger people or property, and to avoid noise sensitive areas such as schools and hospitals. Aircraft should be tested and evaluated for airworthiness and should not be flown more than 400 feet above ground level. If the aircraft is to be flown within 3 miles of an airport, contact with local

* Electronic Code of Federal Regulations: http://ecfr.gpoaccess.gov/cgi/t/text/text-idx?&c=ecfr&tpl=/ecfrbrowse/Title14/14tab_02.tpl.

† AC 91-57.

controlling authorities should be initiated. And, above all, model aircraft should always give way to, or avoid, full-scale aircraft, and observers should be used to assist in that responsibility.*

FAA policy statement AFS-400 UAS Policy 05-01 was issued September 16, 2005, in response to dramatic increases in UAS operations in both the public and private sectors.† The policy was intended to provide guidance to be used by the FAA to determine if unmanned aircraft systems may be allowed to conduct flight operations in the U.S. National Airspace System. AFS-400 personnel are to use this policy guidance when evaluating each application for a certificate of waiver or authorization (COA). Due to the rapid evolution of UAS technology, this policy is to be subject to continuous review and updated when appropriate.‡ The policy was not meant to be a substitute for any regulatory process, and was jointly developed by, and reflected the consensus opinion of, AFS-400, the Flight Technologies and Procedures Division, FAA Flight Standards Service (AFS); AIR 130, the Avionics Systems Branch, FAA Aircraft Certification Service (AIR); and ATO-R, the Office of System Operations and Safety, FAA Air Traffic Organization (ATO).§

The 05-01 policy recognized that if UAS operators were strictly held to the "see and avoid" requirements of 14 CFR Part 91.113, "Right-of-Way Rules," there would be no UA flights in civil airspace.¶

> The right-of-way rule states that "...when weather conditions permit, regardless of whether an operation is conducted under instrument flight rules or visual flight rules, vigilance shall be maintained by each person operating an aircraft so as to see and avoid other aircraft. When a rule of this section gives another aircraft the right-of-way, the pilot shall give way to that aircraft and may not pass over, under, or ahead of it unless well clear."** The FAA's policy supports UA flight activities that can demonstrate that the proposed operations can be conducted at an acceptable level of safety.††

Another collision avoidance rule states that "no person may operate an aircraft so close to another aircraft as to create a collision hazard."‡‡ The FAA also recognizes that a certifiable "detect, sense and avoid" system, an acceptable solution to the see-and-avoid problem for UA, is many years away.§§

Through the implementation of this policy, the FAA has given civil UAS developers and operators two choices: (1) they can operate their systems as public aircraft and apply for a COA that will permit operation of a specific aircraft in a specific operating environment with specific operating parameters and for no more than one year at a time; or (2) they can follow the normal procedures set forth in the Code of

* AC 91-57.
† FAA AFS-400 UAS Policy 05-01, September 16, 2005.
‡ FAA AFS-400 UAS Policy 05-01, September 16, 2005.
§ FAA AFS-400 UAS Policy 05-01, September 16, 2005.
¶ 14 CFR 91.113.
** 14 CFR 91.113(b).
†† FAA AFS-400 UAS Policy 05-01, Supra Note 43.
‡‡ 14 CFR 91.111(a).
§§ FAA AFS-400 UAS Policy 05-01, Supra Note 43.

Federal Regulations to obtain a special airworthiness certificate for their aircraft,* operate the aircraft in strict compliance with all airspace regulations set forth in 14 CFR Part 91, and have them flown by certificated pilots.† The policy also references AC 91-57, Model Aircraft Operating Standards, published in 1981, as it applies to model aircraft, and states that "UA that comply with the guidance in AC 91-57 are considered model aircraft and are not evaluated by the UA criteria in this policy."‡

The FAA has furthermore declared in this policy that it will not accept applications for civil COA, meaning that only military or public aircraft are eligible.§ A *public aircraft* is defined in 14 CFR Part 1.1 as follows:

> *Public aircraft* means any of the following aircraft when not being used for a commercial purpose or to carry an individual other than a crewmember or qualified non-crewmember:
>
> (1) An aircraft used only for the United States Government; an aircraft owned by the Government and operated by any person for purposes related to crew training, equipment development, or demonstration; an aircraft owned and operated by the government of a State, the District of Columbia, or a territory or possession of the United States or a political subdivision of one of these governments; or an aircraft exclusively leased for at least 90 continuous days by the government of a State, the District of Columbia, or a territory or possession of the United States or a political subdivision of one of these governments.
>
> (i) For the sole purpose of determining public aircraft status, *commercial purposes* means the transportation of persons or property for compensation or hire, but does not include the operation of an aircraft by the armed forces for reimbursement when that reimbursement is required by any Federal statute, regulation, or directive, in effect on November 1, 1999, or by one government on behalf of another government under a cost reimbursement agreement if the government on whose behalf the operation is conducted certifies to the Administrator of the Federal Aviation Administration that the operation is necessary to respond to a significant and imminent threat to life or property (including natural resources) and that no service by a private operator is reasonably available to meet the threat.
>
> (ii) For the sole purpose of determining public aircraft status, *governmental function* means an activity undertaken by a government, such as national defense, intelligence missions, firefighting, search and rescue, law enforcement (including transport of prisoners, detainees, and illegal aliens), aeronautical research, or biological or geological resource management.
>
> (iii) For the sole purpose of determining public aircraft status, *qualified non-crew member* means an individual, other than a member of the crew, aboard an aircraft operated by the armed forces or an intelligence agency of the United States Government, or whose presence is required to perform, or is associated with the performance of, a governmental function.
>
> (2) An aircraft owned or operated by the armed forces or chartered to provide transportation to the armed forces if—
>
> (i) The aircraft is operated in accordance with title 10 of the United States Code;

---

* 14 CFR 21.191.

† FAA AFS-400 UAS Policy 05-01, Supra Note 43.

‡ FAA AFS-400 UAS Policy 05-01, Supra Note 43.

§ FAA AFS-400 UAS Policy 05-01, Supra Note 43, § 6.13.

(ii) The aircraft is operated in the performance of a governmental function under title 14, 31, 32, or 50 of the United States Code and the aircraft is not used for commercial purposes; or

(iii) The aircraft is chartered to provide transportation to the armed forces and the Secretary of Defense (or the Secretary of the department in which the Coast Guard is operating) designates the operation of the aircraft as being required in the national interest.

(3) An aircraft owned or operated by the National Guard of a State, the District of Columbia, or any territory or possession of the United States, and that meets the criteria of paragraph (2) of this definition, qualifies as a public aircraft only to the extent that it is operated under the direct control of the Department of Defense.*

In summary, the FAA mandates that one intending to operate an unmanned aircraft in national airspace must do so either under the permission granted by a COA (available only to public entities, which includes law enforcement agencies and other government entities), or with an experimental airworthiness certificate issued pursuant to relevant parts of Title 14 of the Code of Federal Regulations. Specifically proscribed are operations that are of a commercial nature, without the protection of a COA, but ostensibly under the guidelines set forth in AC 91-57.

In recognition that some commercial for-hire UAS operators are flying their systems in national airspace under AC 91-57 guidelines, the FAA published a second policy statement on February 13, 2007.† This notice was a direct response to increasing efforts by U.S. law enforcement agencies and some small UAV manufacturers to introduce systems into operational service on the back of model aircraft regulations. The policy states that the FAA will only permit UAV operations under existing certificate of authorization and experimental aircraft arrangements. The policy states:

> The current FAA policy for UAS operations is that no person may operate a UAS in the National Airspace System without specific authority. For UAS operating as public aircraft the authority is the COA, for UAS operating as civil aircraft the authority is special airworthiness certificates, and for model aircraft the authority is AC 91-57.
>
> The FAA recognizes that people and companies other than modelers might be flying UAS with the mistaken understanding that they are legally operating under the authority of AC 91-57. AC 91-57 only applies to modelers, and thus specifically excludes its use by persons or companies for business purposes.
>
> The FAA has undertaken a safety review that will examine the feasibility of creating a different category of unmanned "vehicles" that may be defined by the operator's visual line of sight and are also small and slow enough to adequately mitigate hazards to other aircraft and persons on the ground. The end product of this analysis may be a new flight authorization instrument similar to AC 91-57, but focused on operations which do not qualify as sport and recreation, but also may not require a certificate of airworthiness. They will, however, require compliance with applicable FAA regulations and guidance developed for this category.

The gap that is created by these policies is a consistent definition of a "model aircraft," and, as discussed in previous sections of this article, some individuals and

* 14 CFR 1.1.

† 72 FR 6689, Supra Note 40.

agencies have taken advantage of this gap to operate small (and not-so-small) UAVs with cameras and other sensing equipment on board, clearly for either a commercial or law enforcement purpose, without having applied for a COA or a special airworthiness certificate.*

3.6 THE FEDERAL AVIATION ADMINISTRATION'S ENFORCEMENT AUTHORITY OVER UNMANNED AIRCRAFT SYSTEMS

The FAA has two issues to face with respect to its enforcement authority over UAS operations. First, it must determine what it *can* regulate, and, second, it must decide what it *will* regulate. The answer to the second challenge largely depends upon a resolution of the first.

The FAA issues six types of regulations: mandatory, prohibitive, conditionally mandatory, conditionally prohibitive, authority or responsibility, and definition/explanation.† Mandatory and prohibitive regulations are enforceable. The other four types represent exceptions or conditions. A thorough analysis of the applicability of a regulation to a particular situation will include answering the following questions: (1) to whom does the regulation apply; (2) what does it say in its entirety; (3) where must the regulation must be complied with; (4) when must it be accomplished; (5) how does it apply to the situation in question; and (6) are there are any special conditions, exceptions, or exclusions.‡

Since unmanned aircraft are "aircraft," and there is no exception found elsewhere in the regulations that excludes UAVs from the definition, one interpretation would be that the FAA has full regulatory authority over all aircraft that are capable of and do fly in the national, navigable airspace. "*Navigable airspace* means airspace at and above the minimum flight altitudes prescribed by or under this chapter, including airspace needed for safe takeoff and landing."§ Minimum safe altitudes are prescribed at 1000 feet above the ground in a congested area, with a lateral separation from objects of 2000 feet, and an altitude of 500 feet above the surface, except over open water or sparsely populated areas. In those cases, the aircraft may not be operated closer than 500 feet to any person, vessel, vehicle, or structure.¶ The exception is when it is necessary for takeoff or landing, in which case the navigable airspace goes to the surface (and along a designated approach path or airport landing pattern).** The 400-foot AGL altitude limit for model aircraft contained in AC 91-57 was probably an observance of the 500-foot minimum safe altitude for manned aircraft operating anywhere except in Class G (uncontrolled) airspace,†† providing a 100 foot "buffer," in addition to the recommendation to not operate within close

* For example, see Web site for Remote Controlled Aerial Photography Association: http://www.rcapa.net/.

† Anthony J. Adamski and Timothy J. Doyle, *Introduction to the Aviation Regulatory Process*, 5th ed. (Plymouth, MI: Hayden-McNeil, 2005), 62.

‡ Adamski and Doyle, Introduction.

§ 14 CFR 1.1.

¶ 14 CFR 91.119.

** 14 CFR 91.119.

†† 14 CFR Part 71.

proximity to an airport. The actual FAA policy history of AC 91-57 is not available for confirmation, but the foregoing is the commonly held belief of FAA officials and individuals familiar with the history of model aviation.*

If the broad definition of *aircraft* is interpreted to include unmanned aircraft, with no exceptions for models, then the FAA may regulate anything and anyone that operates or pilots an aircraft in the navigable airspace. The vast majority of the FARs are intended to provide for safe operations of aircraft that carry people, both for the protection of the crew and passengers, and for people and property on the ground. Although unmanned aircraft have been on the aviation scene for over 90 years, there is no evidence in any of the preambles to regulations or other historical documents currently available for review that the authors of any regulation contemplated application of a specific regulation to unmanned, remotely piloted aircraft. Moored balloons and kites,[†] unmanned rockets,[‡] and unmanned free balloons,[§] categories of objects or vehicles that are intended to occupy a place in the airspace and are unmanned, are specifically covered by existing regulations, but there is nothing similar for other types of unmanned aircraft.

It could be argued that the FAA has some enforcement authority under existing airspace regulations 14 CFR §§ 91.111 and 91.113, which require that an operator of an aircraft be able to safely operate near other aircraft and observe the right-of-way rules, but the more difficult issue is whether such aircraft must meet certification requirements for the systems and the qualification standards, with appropriate certificates, for pilots, sensor operators, mechanics, maintenance personnel, designers, and manufacturers.

As of this publication, there has been no formal legal challenge to the FAA's enforcement authority over unmanned aircraft and their operations. Government contractors, Customs and Border Protection, the U.S. military establishment, and other public aircraft operators have, for the most part, followed the guidelines of AFS-400 UAS Policy 05-01, Interim Operational Approval Guidance 08-01 and AC 91-57. Likewise, there is no anecdotal evidence that the FAA has initiated any enforcement activity against anyone who is, or is perceived to be, operating a UAS outside of these guidelines. Until a robust set of regulations that specifically addresses the unique characteristics of unmanned aircraft is implemented, there is always the chance that someone will fly a commercial UAS in such an open and notorious manner that the FAA is compelled to respond with more than a "friendly" warning letter or telephone call.

The FAA's public position on this issue, as evidenced by the February 13, 2007, policy statement published in the *Federal Register*, is that any unmanned aircraft to be operated in the national airspace, with the exception of radio-controlled models, must comply with the requirements for a COA if it is a public aircraft, or for a special airworthiness certificate if it is a civil aircraft. Thus, for the time being, the agency

---

* Benjamin Trapnell, Assistant Professor, University of North Dakota, Lifetime Member of the Academy of Model Aeronautics.

† 14 CFR 101.11 et seq.

‡ 14 CFR 101.21 et seq.

§ 14 CFR 101.31 et seq.

has answered the second question (what it will regulate) with a broad statement of policy that it is the responsible authority over airspace and aviation.

The next question, then, is even if the FAA exercises its declared authority over airspace and aviation and attempts enforcement against an operator of a "small" (model size) UAS who is using the system for some arguably commercial purpose, without an airworthiness certificate or a licensed pilot in control, just what regulation would be enforced, and what sanction would be appropriate to deter further violations?

There are entrepreneurs and developers around the world whose presence and activities in the civil small UAS market (the UASs are small, the market is not) are putting pressure on the FAA to take the lead in UAS rulemaking. If a farmer or other commercial agriculture concern were to acquire a small system and fly it over fields in what could be characterized as a "sparsely populated" areas, at an altitude where possible conflict with manned aircraft could occur, is there in place a regulatory mechanism to stop this activity? Or, if a commercial photographer were to operate a small UAS equipped with a camera over a similar area for the purpose of photographing the land for advertising or some similar purpose, could the FAA prevent the operation?

The issue for the FAA in the foregoing scenarios is what tools are in the FAA toolbox to enforce whatever regulations it may deem enforceable. These systems do not have an airworthiness certificate. The FAA's central mission is to promote compliance with safety standards.* FAA Order 2150.3A acknowledges that civil aviation depends primarily upon voluntary compliance with regulatory requirements, and only when those efforts have failed should the agency take formal enforcement action.

A certificate holder cannot be deprived of "property" (the certificate) without due process.† Congress has given the FAA authority not only to make the rules,‡ but to enforce them through a number of methods, including issuance of "an order amending, modifying, suspending, or revoking" a pilot's certificate if the public interest so requires.§ Any other certificate issued by the FAA can be "amended, modified, suspended or revoked" in the same manner. The problem with the aforementioned scenarios is that the "pilot" in all likelihood will not be an FAA certificated pilot, because it is not required for such operations, and the aircraft and its systems will not be certified as airworthy, again because it is not required. So long as the operator/pilot does not interfere with the safe operation of a manned aircraft or otherwise enter a controlled airspace (such as in an airport environment) without permission, there may be no violation of any existing regulation.

Taking the scenario a step further, if the pilot/operator inadvertently allows the UAS to come close enough to a manned aircraft to force the latter into an evasive maneuver (not an unlikely event even in a sparsely populated agricultural region), a possible violation of 14 CFR §91.111 (Operating Near Other Aircraft) could ensue. In this situation, the FAA has no certificate to revoke, and thus no statutory or

* FAA Order 2150.3A.

† *Coppenbarger v. FAA*, 558 F. 2d 836, 839 (7th Cir. 1977).

‡ 49 U.S.C. § 44701(a).

§ 49 U.S.C. § 44709 (b); *Garvey v. NTSB and Merrell*, 190 F. 3d 571 (1999).

regulatory authority to proceed with a formal enforcement proceeding pursuant to 49 U.S.C. §44709(b).

This leaves one other mechanism: the civil penalty the administrator may impose against an individual "acting as a pilot, flight engineer, mechanic, or repairman."* The FAA is authorized to assess a civil penalty for violations of certain regulations, up to $400,000 against large entities or companies, and up to $50,000 against individuals and small businesses.† The relevant section of the U.S. Code defines *pilot* as "an individual who holds a pilot certificate issued under Part 61 of title 14, Code of Federal Regulations."‡ Again, an argument could be made that a non-certificate holder would not be subject to even the civil penalty provisions of the U.S. Code, thus leaving the FAA with no effective or realistic enforcement power over "unauthorized" civil unmanned aircraft operations.

3.7 THE WAY FORWARD: THE FUTURE OF UNMANNED AIRCRAFT SYSTEM REGULATIONS

The foregoing discussion suggests that the FAA's enforcement toolbox may be lacking in substance when dealing with ignorant (of existing FAA policy), uncooperative, or openly defiant UAS operators. The day will surely come when the FAA is forced to deal with a UAS operator, pilot, manufacturer, or business entity that is willing to take the FAA to task on its enforcement powers and push the envelope to see how far it can go before a judicial showdown takes place. As market forces create greater opportunities for developers and entrepreneurs to invest capital into more sophisticated systems and bring the industry closer to solving the sense-and-avoid problem, there will be ever-increasing pressure on the FAA to put into place a regulatory structure that will allow the agency to reclaim its "ownership" of the airspace. This necessarily includes implementing reasonable operational and engineering standards through the rulemaking process that will allow the industry to grow while not negatively affecting the overall safety of the aviation environment.

The first task is to define the scope of what the FAA can and should regulate. There must be a definition of *model aircraft* that is precise enough to give notice to the public of the exact nature of the aircraft that will remain unregulated. This definition should include such factors as size, weight, speed, performance capability, and kinetic energy. That would describe the physical attributes of the aircraft and its systems. In addition, there should be a precise description of the locations and altitudes where model aircraft can be flown. If modeling enthusiasts want to create increasingly larger and faster models that could easily overtake and possibly bring down a small general aviation aircraft, they must know where those aircraft can be legally operated and under what conditions.

The civilian UAS community needs to have standards by which admission to the airspace can be assessed and authorized. There must be a workable definition of a "commercial" UAS operation so that there is no confusion about flying a commercial

* 49 U.S.C. §46301 (d)(5)(A).

† 49 U.S.C. §46301 et seq.

‡ 49 U.S.C. §46301 (d)(1)(C).

UAS mission as a model aircraft. A nonenforceable advisory circular such as 91-57 is of little assistance to the FAA as it attempts to deal with commercial, for-hire UAS operators who believe that they are exempt from any certification requirement and understand that advisory circulars are not regulatory and are not rules, nor are FAA policy statements binding on anyone other than the FAA.

The only real alternative for the FAA is to engage in the rulemaking process, subject to the inevitable lengthy comment and revision schedule. That much is clear. What is not clear is how that process should proceed. One approach is simply to amend the current regulations to state that UASs are "aircraft" and that their operators are pilots for all purposes. An exception could be delineated that would exclude the modelers, subjecting everyone else to the full spectrum of Title 14. This approach would require that all UASs be fully certified as airworthy, that their pilots and operators be properly certificated and rated, and that all airspace regulations be fully complied with. The FAA's system of certification is already in place, and all that is lacking are the standards and guidelines that must be met in each applicable category of regulation.

A second approach would be to systematically dissect each and every part and subpart of Title 14 of the CFRs and amend them as necessary, again through the rulemaking process, as required, to incorporate all known characteristics of unmanned aircraft. Many regulations clearly would have no application to UASs (such as those under Part 121 pertaining to passenger seat restraints or flight attendant requirements), while a large portion of the remainder could have application by interpretation, and thus would be candidates for amendment. This process could conceivably take years, but if undertaken, the most logical place to start would be 14 CFR Part 91, Air Traffic and General Operating Rules; Part 71, Airspace; on to Part 61, Pilot and Crewman Certificates; and then to the aircraft design standards found in Parts 21 through 49.

A third alternative would be to create an entirely new part to 14 CFR devoted entirely to UASs, which would incorporate all the issues of "see and avoid" technology, airspace access, pilot qualifications, manufacturing standards, and airworthiness certification.

In the meantime, pending the full integration of UASs into the aviation world, the FAA requires a tool to enforce its authority over the airspace, and to carry out its mandate to promote public safety and to do no harm to the current system through lack of oversight or misguided oversight. This can best be accomplished by a rule that reinforces the FAA's authority over the airspace, and provides for sufficient sanctions against violators who do not possess certificates to be revoked or suspended, or who are otherwise immune from civil penalty.

3.8 CONCLUSION

The aviation environment is complex, dynamic, and littered with pitfalls, landmines, and blind alleys, to mix several metaphors, and the designer, developer, operator, or user of an unmanned aircraft system seeking access to the National Airspace System or in international airspace must proceed with caution to ensure that the rules of engagement are fully understood. The rulemaking and standards development processes for UASs are underway and are sure to be so for the foreseeable future. Active involvement by the industry and the user community in the process is not only encouraged, but is absolutely essential for the industry to grow and evolve in an orderly fashion. The opportunities for technological advancement for unmanned systems, many of which will have a positive impact upon the rest of the world of aviation from a safety and efficiency perspective, are virtually unlimited. The greatest challenge for the FAA and other CAAs around the world is to arrive at coherent, rational, and enforceable policies, procedures, rules, and regulations governing the operation of remotely piloted aircraft, regardless of where they are deployed or for what purpose.

DISCUSSION QUESTIONS

3.1 Discuss the Federal Aviation Act of 1958.

3.2 What is the FAA's "toolbox?"

3.3 List and discuss the three tools the FAA uses to administer the FARs.

3.4 The FAA has supported and sponsored four domestic committees dedicated to developing standards and regulations for the manufacture and operation of unmanned aircraft. List and discuss each committee.

3.5 Discuss the initial intent of Advisory Circular 91-57.

4 Certificate of Authorization Process

Glen Witt and Stephen B. Hottman

CONTENTS

4.1 INTRODUCTION

4.1.1 BACKGROUND

Routine access by unmanned aircraft system (UAS) proponents to the National Airspace System (NAS) is highly desired. The primary user of UASs to date, the Department of Defense (DoD), has described in the most recent unmanned systems roadmap (DoD, 2010) as well as its predecessor roadmaps and in new individual armed services roadmaps the need for NAS access and airspace integration. Although the DoD has significant special-use airspace, future requirements in the NAS dictate growth from 200,000 hours today to an expected 1.1 million hours in the next few years (Weatherington, 2008). Additional federal organizations, such as the departments of Homeland Security, Interior, Energy, and Agriculture; National Aeronautics and Space Administration (NASA); law enforcement; and a variety of other organizations all desire UAS access to the NAS to support their requirements.

The current access to the NAS, outside of special-use airspace, is through the certificate of authorization process or the experimental airworthiness process. This chapter addresses civil airspace for aviation proponents, the history of UAS access to the NAS, and the current certificate of authorization process.

4.1.2 General National Airspace System

Flight safety is achieved globally by all airspace users operating aircraft, including unmanned aircraft, in compliance with established regulatory criteria. Applicable regulations include those created by individual nations for their sovereign domestic airspace (over a nation's land mass and territorial waters to 12 nautical miles from shore) and by the International Civil Aviation Organization (ICAO) for flights operating in international airspace, over the high seas between the territorial boundaries of the world's nations. Flight operating rules developed by the United States are contained in 14 CFR 91 (General Operating and Flight Rules, 2010). ICAO standards are defined in ICAO Annex 2 (Rules of the Air, 2010). The ICAO standards are applicable to all aircraft except state aircraft of a nation and those aircraft involved in military, custom, or police service when mission requirements are not compatible with the ICAO standards. In such instances, it is the responsibility of the state aircraft to have due regard for the safety of other aircraft (Convention on International Civil Aviation, Article 3, 2006).

Access to the NAS and airspace integration involve the interrelationship of (1) various Federal Aviation Regulations (FARs); (2) airspace classification, designation, structure, and alignment; (3) air navigation facilities and airways, jet routes, and area navigation (RNAV) procedures; (4) airports and landing areas; (5) air traffic control (ATC) system's organizational structure and associated ATC operations and procedures; (6) aeronautical charts and flight information publications; and (7) meteorological information.

FARs are the foundation for establishing a safe NAS. There are several FARs that address safety for flight and airworthiness in the NAS. Other FARs establish standards for pilots with respect to qualifications, certification, and medical requirements. In addition, there are separate FARs that determine the operating criteria for all types of aircraft operations. These FARs are effectively interwoven to ensure safety within the NAS.

Knowledge and understanding of airspace classification and special-use airspace (SUA) designation is critical to conducting flight operations that are safe and in compliance with the applicable regulatory criteria. The United States classification of its domestic airspace (14 CFR Part 71, Airspace, 2010) is patterned after the ICAO alphabetical standards (e.g., Class A, B, C, D, E [controlled airspace], and G [uncontrolled airspace], ICAO Annex 11, Appendix 4, 2010) (see Figure 4.1). Within both of these airspace systems, the airspace classification dictates, as appropriate, such factors as flight under visual flight rules (VFRs) and instrument flight rules (IFRs), basic visibility and clearance from clouds minimums, compliance with established traffic patterns and approach procedures for an airport, communication with ATC, equipment requirements, and deviations. SUA is airspace of defined dimensions

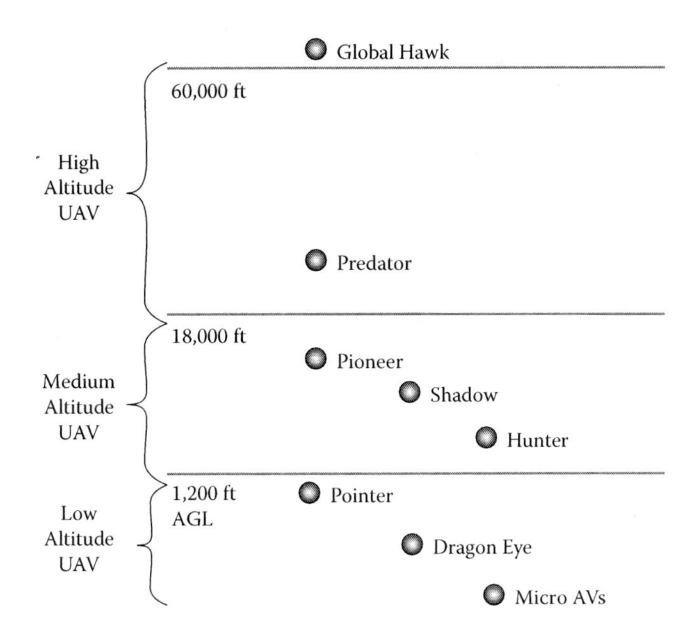

FIGURE 4.1 UASs vary considerably by altitude range.

identified by an area on the surface of the earth where activities must be contained or limited. There are six categories of SUA within the United States NAS: alert area, controlled firing area, military operations area (MOA), prohibited area, restricted area, and warning area. On occasions, as a result of forest fires, natural disasters, and so forth, temporary flight restrictions (TFRs) may be implemented on very short notice through the U.S. Notice to Airmen (NOTAM) system. TFRs also have been applied to selected UAS operations.

The location of air navigation facilities, airways, jet routes, and RNAV procedures are designed to provide capabilities for the flight operator to achieve effective and efficient flight operations. Airports and landing areas are other important components of the NAS. There are a variety of different types of airports and landing areas that provide significant operating capabilities to aircraft operators.

The ATC system, including ATC service providers (control towers, air route traffic control centers, and flight service stations) and the application of ATC procedures are designed to provide services that ensure flight operations in the NAS are performed in a safe, orderly, and expeditious manner. The results of this air traffic management system are reflected in the safety record of manned aircraft operations.

In the future, UAS operators need the ability to perform UAS flight operations in all classes of airspace, eventually on a "file and fly" basis (like a manned aircraft, the pilot files a flight plan, coordinates with the FAA, and then flies soon after that). UAS training operations, border protection "flights," military readiness requirements, and research will affect all airspace classes of the NAS as well as international airspace. It may not be essential that unmanned aircraft operators possess knowledge and expertise in each class of airspace and the knowledge of regulations appropriate to that airspace. Yet it is imperative that unmanned aircraft operators have very thorough knowledge of

FIGURE 4.2 UAS examples.

the regulations applicable to the class of airspace in which the unmanned aircraft flight will be performed. This includes the location of SUA, military training routes, and the process that is used for notification of the activation of SUA and changes to previously published airspace classes, SUA, and other NAS data.

4.1.3 Classes of Unmanned Aircraft Systems

Not all UASs operate in all airspace classes. For instance, a small UAS generally will not operate in Class A airspace due to performance characteristics; therefore, a more limited NAS operating environment would need to be understood. However, a high-altitude, long-endurance UAS would operate in a variety of airspace classes until reaching Class A airspace. In addition to the factors just mentioned, UASs vary greatly in size, weight, power, complexity, autonomy, and their operating altitude (see Figure 4.2). An overview of these classes of UASs described by Deptula (2008) and Weatherington (2008) from the military perspective (although analogous, civil platforms/missions may exist) is as follows:

- Micro/nano UAS—These systems will be used primarily as extremely close-in reconnaissance or communications/network nodes.
- Small—This class of UASs is the hand-launchable, back-packable, quick-look devices, used primarily as an aid to battlefield assessment and situational awareness. The Raven is an existing example of this class.
- Tactical—This class of UASs is used as medium duration, surveillance enhancement systems. They also can perform as communication relays or network nodes. The Shadow 200 is an example of this class.

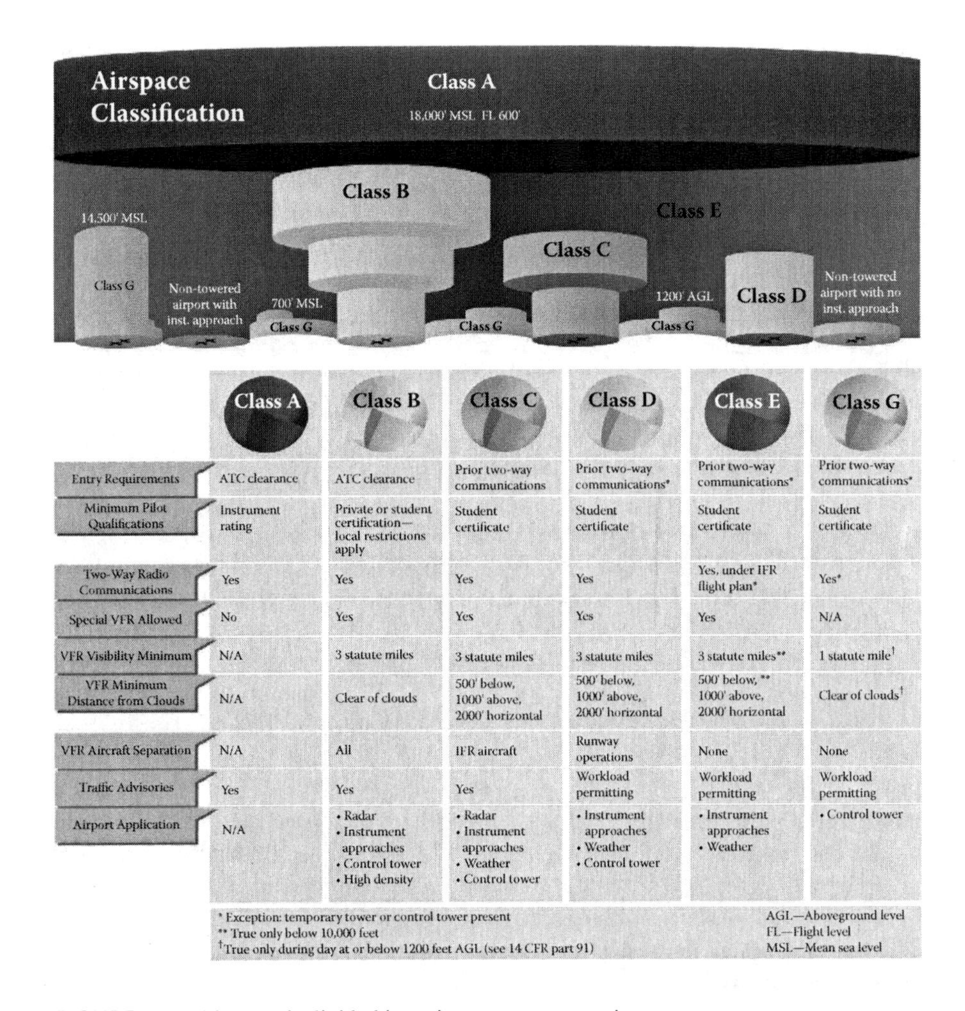

	Class A	Class B	Class C	Class D	Class E	Class G
Entry Requirements	ATC clearance	ATC clearance	Prior two-way communications	Prior two-way communications*	Prior two-way communications*	Prior two-way communications*
Minimum Pilot Qualifications	Instrument rating	Private or student certification—local restrictions apply	Student certificate	Student certificate	Student certificate	Student certificate
Two-Way Radio Communications	Yes	Yes	Yes	Yes	Yes, under IFR flight plan*	Yes*
Special VFR Allowed	No	Yes	Yes	Yes	Yes	N/A
VFR Visibility Minimum	N/A	3 statute miles	3 statute miles	3 statute miles	3 statute miles**	1 statute mile†
VFR Minimum Distance from Clouds	N/A	Clear of clouds	500' below, 1000' above, 2000' horizontal	500' below, 1000' above, 2000' horizontal	500' below, ** 1000' above, 2000' horizontal	Clear of clouds†
VFR Aircraft Separation	N/A	All	IFR aircraft	Runway operations	None	None
Traffic Advisories	Yes	Yes	Yes	Workload permitting	Workload permitting	Workload permitting
Airport Application	N/A	• Radar • Instrument approaches • Control tower • High density	• Radar • Instrument approaches • Weather • Control tower	• Instrument approaches • Weather • Control tower	• Instrument approaches • Weather	• Control tower

* Exception: temporary tower or control tower present
** True only below 10,000 feet
† True only during day at or below 1200 feet AGL (see 14 CFR part 91)

AGL—Aboveground level
FL—Flight level
MSL—Mean sea level

FIGURE 4.3 Airspace is divided into six separate categories.

- Medium—This class of UASs is used for intelligence, surveillance, and reconnaissance (ISR) capability, and as a system to deliver lethal munitions; fulfilling the role of a hunter/killer. The Sky Warrior is an example of this class of UASs.
- Strategic—This class of UASs is used as a strategic asset, with near global reach. It typically is used for long-duration surveillance, with capability typically found in orbiting satellites. An example is Global Hawk.
- Special—This class of UASs will be the next generation of assets deployed to act as surrogate fighter/attack aircraft, as well as airlift capability.

The variety of UASs (see Figure 4.3) and their subsequent use suggests that not all airspace knowledge be the same. The operator of a UAS operating within a few hundred feet above ground level (AGL) may require less knowledge, skills, and ability

(KSA) than the operator of a large, more robust system in higher airspace with other unmanned and manned platforms.

4.2 UNMANNED AIRCRAFT SYSTEM AIRSPACE ACCESS HISTORY

The FAA's method for authorizing unmanned aircraft flight operations in the NAS has varied since its initial authorization criteria was established (Hottman, Gutman, and Witt, 2000). Since the DoD was and continues to be the primary user of UASs, the initial FAA criteria and approval for the operation of unmanned aircraft in the NAS addressed these DoD concerns. This criterion was developed in the early 1980s and was contained in the joint FAA and DoD Order 7610.4, Special Military Operations (2001). At that time, the FAA classified unmanned aircraft as remotely piloted vehicles (RPVs). There were few unmanned flight operations at that time and the actual number of unmanned aircraft in existence was very small.

When the FAA established the criterion for DoD unmanned aircraft operations, no specific requirements for operation of unmanned aircraft by other public or civil organizations were developed and published (Hottman & Copeland, 2005). When these non-DoD organizations began to request approval to operate unmanned aircraft in the NAS, FAA headquarters instructed their regional offices to apply the same criteria that were being used for DoD unmanned aircraft.

The initial FAA criterion for the DoD provided unlimited access to DoD unmanned aircraft flights in the NAS that were flown within restricted areas, warning areas, and positive control areas (PCAs). PCA was basically what is now termed as Class A airspace. Flight operations in PCAs and their successor, Class A airspace, required all aircraft, including unmanned aircraft, to be operated under IFRs and in compliance with an ATC clearance. This condition creates a positive control environment so that safety is maintained through ATC providing separation between all aircraft, including unmanned aircraft operating under IFRs in that airspace.

The FAA criterion also authorized the DoD to operate unmanned aircraft outside of restricted areas, warning areas, and below PCA/Class A airspace in other classes of airspace of the NAS without having to obtain additional FAA approval, provided a chase aircraft accompanied the unmanned aircraft during the flight. The early FAA criterion also provided that in some instances, where no chase aircraft would be used, the FAA could separately approve DoD unmanned aircraft flights in the NAS. This included outside of restricted areas, warning areas, and PCA/Class A airspace. In these latter instances, the DoD was required to provide an alternate method of observing the unmanned aircraft other than by a chase aircraft (e.g., patrol aircraft, radar monitoring, ground observers, or the use of the controlled firing area concept). The DoD's proposed use of any of these alternative methods was evaluated by the FAA for each specific unmanned aircraft operation to determine if the proposed alternate method could provide an equivalent level of safety as that provided using a chase aircraft before an approval was granted.

In order to provide an approval authority for all unmanned aircraft flights in the NAS that required additional FAA approval, FAA Order 7610.4 (Special Military Operations, 2001) delegated authority to each of the FAA's nine regions individually. The responsibility for making the determination was decided by the FAA region that was responsible for managing the airspace where the flight originated. If the airspace for the unmanned aircraft flight operation overlapped FAA regional boundaries, the FAA region responsible for the determination would coordinate with the other affected FAA region(s) before approving the unmanned aircraft flight activity.

4.2.1 Federal Aviation Administration Memorandum Direction

The initial criterion the FAA applied to proponents (primarily DoD) requesting to operate unmanned aircraft in the NAS remained unchanged until 1992 (Witt and Hottman, 2006). Leading up to 1992, the number of requests to operate unmanned aircraft in the NAS increased. In a September 1992 memorandum to its regional air traffic and flight standards division managers, FAA headquarters issued a policy statement concerning the handling of unmanned aircraft operations. In this joint policy statement, the associate administrator for air traffic and the associate administrator for regulation and certification indicated that the FAA was receiving requests from unmanned aircraft proponents in increasing numbers. The FAA associate administrators also indicated the requests varied from the need to test unmanned aircraft that are under development to actual applications such as drug interdiction, upper air sampling, aerial photography, and border patrol.

The heightened concern of the FAA in 1992 regarding unmanned aircraft flight operations in the NAS centered on the fact that except for 14 CFR Part 91 (Rules of the Air, 2010), there were no specific federal aviation regulations governing the operation of an unmanned aircraft. The FAA believed that because of the uncertainty regarding the extent of FAR Part 91, unknown, non-DoD organizations were operating unmanned aircraft in the NAS. Therefore, the goal of the joint memorandum was to specify how the FAA regions would process requests by unmanned aircraft proponents, both DoD and non-DoD organizations.

The FAA's 1992 policy statement reiterated that requests from the DoD to conduct unmanned aircraft operations in the NAS would continue to be processed under the procedures of FAA Order 7610.4 (Special Military Operations, 2001). However, this new policy established the Air Traffic Rules and Procedures Service in the FAA Washington headquarters as the office of primary interest (OPI) for all non-DoD unmanned aircraft operations. The policy statement went on to clarify that in order to ensure standard application of this policy, regions may only approve a non-DoD unmanned aircraft operation in a prohibited area, restricted area, or warning area. The policy statement also stated that all other requests were first to be studied by the appropriate regional division toward arriving at a recommended disposition of the request and then were to be forwarded to the OPI for further processing. While the 1992 policy statement clarified which FAA organizations were responsible for processing and approving DoD or non-DoD requests for unmanned aircraft operations, there were no guidelines on what materials the unmanned aircraft proponent was required to submit to the FAA. Nor were there guidelines regarding the document the FAA would use to approve an unmanned flight operation and define any FAA requirements as part of the approval.

4.2.2 A New Airspace Forcing Function

Technology became the next forcing function for a change to airspace access. In 1995, the advent of the United States Air Force's Predator A unmanned aircraft and its enhanced command and control technology increased the flight ability of unmanned aircraft from local flight operations to those regional in scope. Similarly, the introduction in 1998 of the Air Force's Global Hawk unmanned aircraft with satellite communication command and control enabled international unmanned aircraft flight operations for the first time. During the development of the Predator A and Global Hawk unmanned aircraft, there was considerable discussion within the FAA concerning what procedures would be necessary to integrate these unmanned aircraft flight operations safely into the NAS.

The initial outcome of the FAA discussion on the procedures for ensuring safety while enabling more DoD unmanned aircraft flights in the NAS was to change the FAA's criterion for approving DoD unmanned aircraft flights in the NAS. In March 1999, the FAA, through its internal notice process, issued a Notice (N7610.71):

> Subject: Department of Defense (DoD) Remotely Operated Aircraft (ROA) Operations (1999). This notice implemented a change to FAA Order 7610.4, Special Military Operations, Chapter 12, Section 9, Remotely Piloted Vehicle (Special Military Operations, 2001).

As part of this change the FAA reclassified unmanned aircraft from remotely piloted vehicle (RPV) to remotely operated aircraft. The most significant change was that, except for unmanned aircraft flights in restricted areas and warning areas, all other DoD unmanned aircraft would be required to have specific FAA authorization prior to operating in the NAS. DoD unmanned aircraft organizations no longer were authorized to operate their unmanned aircraft in Class A airspace or in other airspace using a chase aircraft without having to obtain specific FAA approval. Also, for the first time, the FAA published the details of what information and data DoD unmanned aircraft proponents had to submit to the FAA when making their request.

4.3 CERTIFICATION OF AUTHORIZATION (COA) OR WAIVER IMPLEMENTATION

Part of the changes the FAA implemented in 1999 established the use of the FAA's certificate of authorization or waiver as the process that DoD organizations were to use in making their request to perform unmanned aircraft flight operations in the NAS, outside of restricted or warning areas. The guidelines for the certificate of waiver or authorization process are contained in FAA Order 7210.3, Facility Operation and Administration, Part 6, Chapter 18, Waivers, Authorizations, Exemptions, and Flight Restrictions (2010). This process for unmanned aircraft became commonly known as the certificate of authorization or COA process. As before, the FAA did not publish any specific guidelines for non-DoD unmanned aircraft proponents to use, so organizations merely used the previously defined criterion for the DoD.

The 1999 FAA criteria required all DoD unmanned aircraft proponents who desired to conduct unmanned aircraft flight operations in the NAS, outside of restricted and warning areas, to submit an application for a COA at least 60 days prior to the planned commencement date of the unmanned aircraft operation. The application for COA was to be sent to the Air Traffic Division of the appropriate FAA regional office. For the first time, the FAA specified what information and data the unmanned aircraft proponent was to submit to the FAA, which became part of the COA application. The following data and information were required:

1. Detailed description of the intended flight operation, including the classification of the airspace to be used
2. Unmanned aircraft's physical characteristics (configuration, length, wingspan, gross weight, means of propulsion, fuel capacity, color, lighting, etc.)
3. Flight performance characteristics (top speed, cruise speed, maximum altitude, rate of climb, range/endurance, means of recovery, etc.)
4. Method of pilotage and proposed method to avoid other air traffic
5. Coordination procedures
6. Communication procedures
7. Route and altitude procedures
8. Lost link/mission aborts procedures
9. A statement from the DoD proponent that the unmanned aircraft is airworthy

The air traffic organization is a unit of the FAA, which continued to be responsible for approving unmanned aircraft flight operations in the NAS. The air traffic organization coordinates with certain FAA regulatory and safety organizations. However, for DoD unmanned aircraft operations, one of the FAA's nine regional offices in the Air Traffic Division was the approval organization. Published criteria defining unmanned aircraft equipment and operating capability requirements did not exist. Therefore, there were inconsistencies between what one region would authorize and another would not authorize. In February 2004, the FAA changed its air traffic organizational structure from nine regional offices to three service areas. The three service areas now were responsible for making the determination of whether, in their opinion, a particular DoD unmanned flight operation would be allowed to operate outside restricted or warning area airspace in the NAS. The service area organizations had an increased understanding of unmanned aircraft technology and operational capability. As communication increased among the service areas, approving DoD unmanned aircraft flight operations became more standardized.

The U.S. Air Force's Global Hawk unmanned aircraft demonstrated the extent and capabilities of unmanned aircraft. As part of the Air Force's development of this unmanned aircraft technology, Global Hawk made international flights to Australia in April 2001 and Germany in 2003 (Hottman and Witt, 2006). During this period, the FAA continued its effort in developing processes and procedures to ensure the safety of each unmanned aircraft flight operation that was flown outside restricted and warning areas. The FAA's primary concern was that the flights did not adversely impact manned aircraft airspace users' flight operation. In addition, the flight of the Altair UAS from California to Alaska in 2004 illustrated the coordination involved with a non-DoD unmanned high-altitude long endurance (HALE) aircraft flight. The Altair flight traversed domestic airspace, offshore airspace and oceanic airspace, special-use airspace (warning areas), and the ADIZ (air defense identification zone) boundaries of two nations (14 separate airspace owners or managers required coordination). The advanced coordination that was required for this flight operation exemplified the effort required in planning and accomplishing longer duration or distance missions (Hottman and Witt 2006; Witt and Hottman 2006).

4.4 FEDERAL AVIATION ADMINISTRATION GUIDANCE DOCUMENTATION

The next evolution in the FAA's effort to ensure that unmanned aircraft flights in the NAS could be conducted safely was the issuance of a memorandum by the FAA's Flight Technologies and Procedures Division, AFS-400 in September 2005. This memorandum defined the FAA's "Interim Operational Approval Guidance" and established a new terminology for unmanned aircraft as unmanned aircraft system (UAS) to replace the term of remotely operated aircraft. The memorandum stipulated that AFS-400 personnel would use this policy guidance when evaluating each application for a COA. Also, for the first time, the FAA specified that "civil" unmanned aircraft proponents would not be able to use the COA process and should follow the FAA's current airworthiness certification processes. The AFS-400 policy statement also specified that applications for COAs submitted by public unmanned aircraft proponents must include one of the following:

- A civil airworthiness certification from the FAA
- A statement specifying that the Department of Defense Handbook "Airworthiness Certification Criteria" (Military Handbook 516) was used to certify the aircraft
- Specific information explaining how an airworthiness determination was made

The AFS-400 memorandum also, for the first time, established specific equipment, operational, and personnel requirements for unmanned aircraft that were to be flown in the NAS. These requirements included criteria for:

- Chase aircraft operations
- Communication between flight crew personnel
- Flight operations
    - Within Class A airspace
    - Within Class C, D, E, and G airspace
    - Flight over congested or populated areas
    - Lost link
- Observer qualifications
- Pilot qualifications
- Pilot/observer medical standards

- Pilot responsibilities
- Radar/sensor observers
- Visual observer responsibilities

4.5 CREATION OF UNMANNED AIRCRAFT PROGRAM OFFICE

The FAA took a major step toward resolving some of the uncertainties of unmanned aircraft operations when it created its Unmanned Aircraft Program Office (UAPO) in February of 2006. The purpose of the UAPO is to develop policies and regulations that ensure unmanned aircraft operate safely in the NAS. The UAPO is comprised of individuals with FAA safety, regulatory, engineering, and air traffic service experience.

4.5.1 Certificate of Authorization Focus

One of the first tasks of the UAPO was to develop a comprehensive COA application process that ensured the FAA was receiving sufficient information and data from public unmanned aircraft proponents. The comprehensive COA application process determines if the unmanned aircraft was airworthy and that the flight operations did not pose a hazard to other airspace users or persons on the surface. Initially, the UAPO required the public unmanned aircraft proponent to provide the same data that had been established by the AFS-400 policy in 2005. However, the UAPO quickly undertook the task of modifying what data the public proponent would submit in the COA application. The objective of modifying the COA application requirement was for the FAA to obtain more detailed information and data to enhance the FAA's knowledge of the unmanned aircraft proponent's unmanned aircraft system technology, proposed flight operations, and the qualifications and aviation medical status of the individuals involved in the flight operations of the unmanned aircraft. A second feature of the enhanced COA application was for the FAA to obtain data and information that could be used to develop an extensive database of unmanned aircraft technology, capabilities, and personnel qualifications that could be used in the future to develop regulatory criteria.

4.5.2 COA Application Data and Steps

The enhanced COA Application requires the public unmanned aircraft proponent to provide details described in Table 4.1. Initially the COA application process was performed manually; however, during 2007, the UAPO's Air Traffic Organization worked diligently to create a Web-based COA application system. In late November 2007, New Mexico State University's (NMSU) Unmanned Aircraft Program was the first public unmanned aircraft organization to submit an application for COA via the COA online system. The online COA system now is available to all public unmanned aircraft proponents to use. The COA online system has significantly reduced the workload and amount of time it takes the public unmanned aircraft proponent to develop the COA application and submit it to the FAA. With the COA

TABLE 4.1
Required Information for the COA Application

Item	Description
Proponent information	Identifies the organization and individual from the organization
Point of contact information	Identifies the individual that is the point of contact between the applicant and the FAA
Operational description	Specifies the proposed beginning and ending dates; briefly describes the overall program objectives; specifies whether flight operations will be performed with lights out, under VFR, and/or IFR; and day and/or night operations; identifies the location (state, county, and nearest airport); indicates what class of airspace the flight operations will be performed in (A, B, C, D, E, and/or G); also includes an operation summary section that includes information that is not requested elsewhere in the application
System description	Identifies the unmanned aircraft; control station (number of stations, remote control, etc.) and communication systems description; files normally are attached that provide this information, including photos
Certified technical standard order	Components or other system information
Performance characteristics	Climb rate (feet per minute [fpm]); descent rate (fpm); Turn Rate (degrees per second); cruise speed (knots indicated airspeed, maximum/minimum); operating altitudes (maximum/minimum mean sea level [MSL] or flight level); approach speed; gross takeoff weight (lbs); launch/recovery (description, type/procedures)
Airworthiness	FAA type certificate or a statement on official organization's letterhead stationery regarding the activity that has been conducted by the proponent to validate the airworthiness qualities of the unmanned aircraft
Procedures	Specifies the procedures that will be used for lost/link, lost communication, and emergency situations
Avionics/equipment	Lists the transponder suffix and specifies whether the unmanned aircraft's equipment includes GPS, moving map indicator, tracking capability, terminal control area (TCA) midair collision avoidance system (MCAS), emergency locator transmitter (ELT), and transponder. In addition, the proponent identifies the transponder capabilities (on/off, standby, ident, mode S, mode C, transponder retuneable in flight)
Lights	Indicates whether the unmanned aircraft lights include landing, position/navigation; anticollision, and infrared (IR)
Spectrum analysis approval	Indicates whether spectrum analysis and approval has been obtained for the data link and the control link(s) and provides any approval document; additionally specifies whether any operations will utilize radio control (RC) frequencies as described in Title 47 CFR 95

(continued)

TABLE 4.1 (continued)
Required Information for the COA Application

Item	Description
ATC communications	Specifies the two-way voice capability (instantaneous), whether is a VHF, UHF, HF transmitter and receiver and guard (emergency) frequencies; in addition, the proponent indicates whether instantaneous two-way communication exists for direct to pilot, SATCOM, and relay via the unmanned aircraft
Electronic surveillance/detection capability	Indicates whether the onboard equipment consists of electro-optical/infrared, terrain detection, weather/icing detection, radar, or electronic detection systems; if electronics detection systems are onboard, description is provided; in addition, the proponent indicates whether radar observation (ATC, etc.) will exist
Aircraft performance recording	Indicates whether flight data record, control station recording, and voice recording is available
Flight operations area/plan	Proposed flight operations area(s) may be defined by latitude/longitude points or by specifying a nautical mile radius of a single latitude/longitude point. Identify the altitude for the floor and ceiling of each defined area and the minimum and maximum speed that the unmanned aircraft will operate; in addition, a map must be provided that depicts each area where flight operations are planned
Flight aircrew qualifications	Identifies FAA or DoD equivalent for all pilots and observers: private (written); private (certified); instrument; commercial; air transport; unique trained pilot; and a description of each; provides records for any of the aircrew that are DoD certified/trained; indicates the medical rating, describes the currency status, and duty time restrictions for each crew member; indicates whether only a single unmanned aircraft will be controlled and provides a description of how the unmanned aircraft will be controlled. If the proposed number of unmanned aircraft that are to be controlled simultaneously is more than one, the proponent must state what number of unmanned aircraft will be controlled simultaneously and flight medical classification (FAA or DoD equivalent) for each flight crew member
Special circumstances	May include information or data that the proponent believes is significant and has not been provided previously in this COA application

online system, applicants may establish an account, submit a COA draft, and follow the process online.

Submitting a COA online is a several step process that can be done incrementally. First, applicants need to establish an account by contacting the FAA's UAPO Air Traffic Organization (ATO). After an account has been established, the applicant may go online and initiate a draft, which is then assigned a UAS COA case number. From that point, the applicant can begin populating the various sections of the application with the necessary information and data. The draft does not have to be completed at a single session. The applicant can save the information and return to it at a later time to insert additional information and data. Once all of the necessary information and data has been inserted into the draft, the applicant uses the "commit case" feature to submit the COA application. The system automatically performs an audit to determine that every section of the draft has been filled in. If the system recognizes that information or data has been inserted in every required area, the applicant will receive an "accept" message.

After the applicant has electronically submitted a completed draft, the draft is reviewed in several steps. First, the FAA's UAPO ATO determines whether the information and data are sufficient for the next step, comprehensive analysis. If this precursory review reveals that there is adequate information and data, the applicant will receive a message from the FAA that the COA application submitted has been validated. If the precursory review reveals additional information or data are needed, the applicant will receive a message that the COA case has been "released" back to the applicant. The area where more information is needed by the FAA will be specified, and the applicant has control of the COA case until it is resubmitted. Once the COA case is validated, the 60-day approval time period begins.

After the COA case is validated, the FAA's comprehensive review and analysis process begins. The applicant may be contacted for additional information or data. Once the FAA has probable approval of the COA application, the UAPO ATO is contacted. The UAPO ATO initiates coordination with the appropriate FAA service area offices, which in turn coordinate with any ATC facility that controls the airspace where the flight will be performed. When all appropriate FAA UAPO elements agree that the proposed unmanned aircraft flight operation can be performed safely in the NAS, the COA is issued to the applicant organization.

The COA generally covers a one-year period for flight operations. Public unmanned aircraft organizations may want to continue the flight operation of their unmanned aircraft beyond the one-year period specified. When the unmanned aircraft organization begins to develop continued flight operations beyond the effective period stated in the existing COA, the applicant can duplicate all of the information and data that were submitted in the previous COA application (clone this case) and start a new draft. The applicant can modify the content and bring the information up to date. Once the appropriate changes have been made, the new COA case is submitted, and the FAA processing procedures start again. The FAA continually works to simplify this online process in response to an increase in COA applications.

4.6 FUTURE UAS NATIONAL AIRSPACE SYSTEM ACCESS ADVANCES

The FAA processes for UAS access to the NAS have evolved over the last several decades. A variety of proponent organizations, such as the DoD and Department of Homeland Security, have advocated significantly for greater NAS access. Constituent groups, such as RTCA, AUVSI, ACCESS 5, UAS TAAC, and others, have constructively endeavored to facilitate opening the airspace by soliciting the FAA and

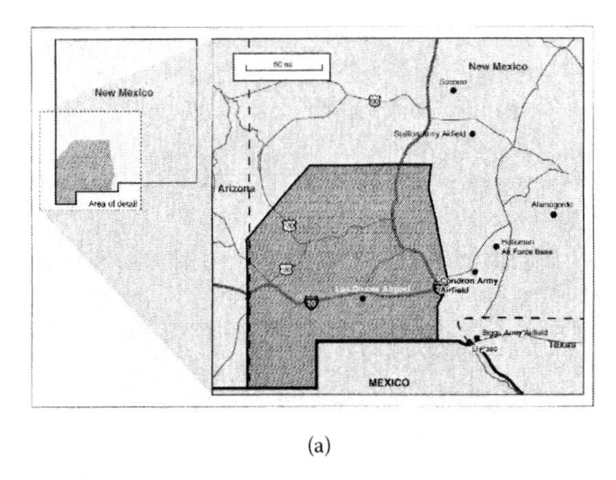

(a)

(b)

FIGURE 4.4 UAS airspace associated with the Flight Test Center and an example of an operating UAS.

coordinating with other proponents (Hottman, Hansen, Sortland, & Wernle, 2004; Timmerman, 2005).

The lack of a UAS-specific, or applicable, regulatory framework has been recognized by the FAA (Hickey, 2007) and other organizations. One early UAS regulatory related effort was the HALE Roadmap (Nakagawa et al., 2001) developed as part of the Environmental Research Aircraft and Sensor Technology (ERAST) program. In 2008, the Center for General Aviation Research, through the efforts of the University of North Dakota, completed a review of FAR applicability to UASs. An accepted regulatory body is necessary to advance UAS access to the NAS (Hottman, 2008). The FAA recently has entered into a cooperative research and development agreement with NMSU creating a UAS Flight Test Center in the NAS with the benefit to the FAA being the receipt of all empirical data from the center, which, in turn, supports the UAS regulatory development (GAO, 2008). (See Figure 4.4.) In recent years, a small UAS rule-making committee was convened by the FAA to propose a federal regulation for these aircraft. The recommendation has been received from the committee and the rule is in process as of 2011.

The DoD continues to be the largest user of UASs domestically and internationally, while the DHS has significant domestic operations, and NASA has a growing science mission with its Global Hawk. During 2009, a new initiative was created to facilitate coordination and advocate for more progress with NAS access for UAS. This initiative was created in the form of an executive committee (ExCom) comprised of senior leadership from the FAA, DoD, DHS, and NASA. The ExCom is active with goals established and work products in progress primarily related to airspace access and integration for UAS in the NAS.

Finally, the system and technology aspects related to UAS developments must not be ignored. ATM has been developed for the safety of all NAS users. The COA process for UASs is part of air traffic management and control of the NAS. Technology advances for UASs are being worked now, ranging from detect, sense, and avoid; automation; communication; and the expectation that next-generation technologies also will positively impact UASs. As the FAA continues to mature its database on UASs and as technology for UASs matures, one should expect that the COA-type process will, in the future, result in NAS UAS operations that have reduced limiting conditions, leading toward a file-and-fly environment desired by many proponents.

4.7 CONCLUSION

The FAA's processes for approving both public and civil unmanned aircraft flight operations in the NAS have continued to evolve over the past three decades. The FAA is continuing to improve these processes and is making it easier for unmanned aircraft organizations to submit their requests to fly unmanned aircraft in the NAS. Some unmanned aircraft proponents may think that the FAA COA and airworthiness certification processes for unmanned aircraft are too cumbersome and complex. Yet, if unmanned aircraft were held strictly to the manned aircraft standards specified in the FARs, there would be no unmanned aircraft flights in the NAS outside of prohibited, restricted, or warning areas. The COA and aircraft airworthiness processes for unmanned aircraft are mechanisms for maintaining aviation safety and must be seen as such.

The COA and airworthiness certification processes are integral parts of the FAA's effort to enable both public and civil unmanned aircraft to be flown in the NAS safely, thus furthering the advancement of this relatively new aviation technology. Currently, the vast majority of unmanned aircraft are owned and operated by public organizations. Therefore, the COA process is used to a much greater extent than the airworthiness certification process. The COA process is basically composed of three major steps:

1. Development and submission of the COA application by the public unmanned aircraft proponent.
2. FAA's comprehensive review and analysis of the information and data submitted in the COA application in order to:

a. Confirm the unmanned aircraft's airworthiness status and the equipment components and associated technology are sufficient and reliable enough to support safe flight operations.
b. Perform coordination with the appropriate FAA service area and affected ATC field facilities to verify the proposed flight area does not adversely impact the operations of other airspace users, is not situated over any populated surface area, and the actual flight operations are compatible with the ATC system in that location.
c. Develop the special provisions that will be included in the COA to clearly define the operating limitations and procedures to be used for the flight operations.

3. Issuance of the COA for a period to not exceed one year in duration.

REFERENCES

Airspace. (2010). 14 CFR Pt. 71 (June 10, 2010).

Convention on International Civil Aviation. (2006). Article 3, International Civil Aviation Organization. Doc 7300/9, 9th Ed. Montreal, Quebec, Canada.

Department of Defense. (2009). Unmanned Systems Integrated Roadmap 2009–2034. Washington, D.C.: DoD.

Deptula, D. (2008, December). The indivisibility of intelligence, surveillance, & reconnaissance (ISR). UAS TAAC 2008 Conference, Albuquerque, NM.

FAA. (2008). Unmanned aircraft systems operations in the U.S. national airspace system. Interim Operational Approval Guidance 08-01.

FAA Notice (N7610.71), Subject: Department of Defense (DOD) Remotely Operated Aircraft (ROA) Operations (1999).

Facility Operation and Administration, Part 6, Chapter 18, Waivers, Authorizations, Exemptions, and Flight Restrictions. FAA Order 7210.3, February 11, 2010.

Government Accountability Office. (2008). Unmanned aircraft systems: Federal actions needed to ensure safety and expand their potential uses with the national airspace system (Technical report GAO-08-511).

Hickey, J. (2007). FAA's program plan for unmanned aircraft: Impact of shifting priorities. Presented to the AIA Subcommittee on UAS.

Hottman, S. (2008, February 28). RDT&E to advance UAS access to the NAS. Briefing for NDIA, Palm Springs, CA.

Hottman, S. B., and Witt, G. (2006). Creation of special use airspace for unmanned aircraft activity: A concept for creating greater opportunities for unmanned aircraft flight operations. Presented at the AUVSI Conference, August 30, 2006.

Hottman, S. B., and Copeland, P. A. (2005). A Systems Approach to Introducing UAVs into Civil Airspace Operations. *UAVNET*.

Hottman, S., Hansen, K., Sortland, K., and Wernle, K. (2004). UAV Operating Procedures in the National Airspace System: Preliminary Findings from a New DOD Program. Proceedings for the Association for Unmanned Vehicle Systems International.

Hottman, S., Gutman, W., and Witt, G. (2000). *Research and validation of requirements for UAV operations in the USA*. Paris, France: UAV.

ICAO. (2010). ICAO Annex 2, Rules of the Air. March 2010.

ICAO. (2010). ICAO Annex 11, Air Traffic Services. March 2010.

Nakagawa, G., Witt, G., Hansen, K., & Hottman, S. B. (2001). HALE UAV Certification and Regulatory Roadmap. TAAC Document C020-01.

Rules of the Air. (2010). 14 CFR Pt. 91 (June 10, 2010).

Special Military Operations. (2001). FAA and DoD Order 7610.4. (July 12, 2001).

Timmerman, J. (2005). Authorizing operations in the National Airspace System. UAS TAAC 2005 Conference, Albuquerque, New Mexico.

Weatherington, D. (2008, December). Unmanned aircraft systems. *UAS TAAC 2008 Conference,* Albuquerque, NM.

Witt, G., & Hottman, S. B. (2006, March 29–30). Human and airspace factors affecting UAV Operations. UV Asia–Pacific 2006 Conference, Sydney, Australia.

5 Unmanned Aircraft System Operations

Theodore Beneigh

CONTENTS

5.1 OBSTACLES TO UNMANNED AIRCRAFT SYSTEM OPERATIONS

Until the mid-1990s, the use of unmanned aircraft systems (UASs) has been, for the most part, restricted to military operations. The use of aircraft in a military combat theater is significantly different than during peacetime operations. The acceptability of risk is significantly higher, and there are very little, if any, nonmilitary flight operations in a combat theater. Hence, the rules and procedures for separating aircraft from collision as well as the threats participating aircraft pose to persons and objects on the ground are radically different than peacetime operations.

Prior to the mid-1990s, nearly all "unmanned aircraft" were model aircraft. The term *model* can be quite deceiving; some of these models are quite large and capable of speeds in excess of 200 knots. On June 6, 1981, the Federal Aviation Administration (FAA) published an advisory circular—designated AC 91-57—defining model aircraft and creating their operating standards. This AC states that model aircraft cannot be flown higher than 400 feet above the ground, or within 3 miles of an airport, unless the authorities in control of the manned aircraft at that airport are notified. Also, these models should be flown away from noise sensitive areas, such as schools, hospitals, and churches. The purpose of these restrictions is to eliminate any collision possibility with manned aircraft and to protect the population from injury caused by a crash of a model aircraft.

Beginning in the mid-1990s, unmanned aircraft began to prove their worth in the military combat areas in the Middle East. Initially, they were used successfully for an array of reconnaissance missions. This success was noted by various nonmilitary agencies, and applications were developed that would allow the unmanned aircraft to perform the mission currently being performed by manned aircraft at a fraction of the cost. In addition, their small size would allow them to undertake missions not possible by the larger manned aircraft. Subsequently, in recent years it has fallen upon the FAA along with other organizations to determine how to integrate the UA (unmanned aircraft refers soley to the air vehicle) into the National Airspace System (NAS). The primary goal of the FAA is to regulate and oversee all aspects of civilian aviation in the United States, with the emphasis on safety.

The key component of collision avoidance between aircraft is the ability for each pilot to see and avoid each other using visual acquisition. When weather conditions are too poor for visual avoidance, instrument flight rules (IFR) are in effect. During IFR flight, collision avoidance is provided by the FAA through the use of air traffic control. Since there is no pilot onboard a UA, the ability to see and avoid other manned aircraft is compromised.

Manned aircraft have very rigid standards to which they must be built and certified. These are defined by the Code of Federal Regulations (CFR) Parts 23 and 25. Essentially, these regulations provide assurances to those who fly in them, and those over which they fly, that the aircraft will remain intact and safely perform the flight operation for which it is certified. UASs, on the other hand, have little government regulation. The pilot, not being onboard the UA during flight, can easily put the UA through a maneuver that can compromise its structural integrity, which could cause an inflight breakup. This would pose a significant hazard to people on the ground, below.

It is possible that the pilot on the ground can lose communications with the UA during flight. If this happens, where will the UA go? Foolproof procedures must exist that would preclude the UA from presenting a hazard should this occur.

5.2 GUIDELINES TO UAS OPERATIONS

Recognizing the new problems the operation of UAs in the NAS would create, the FAA has created a guidance document: FAA Interim Operational Approval Guidance 08-01. Although this document is not regulatory, it dictates the procedures that are being used by the FAA to allow UAs access to the NAS. It is dynamic and is changed as FAA policy changes. FAA Interim Operational Approval Guidance 08-01 identifies alternate methods of compliance with the regulations when evaluating proposed UAS operations. It was developed to mitigate the obstacles listed earlier and to provide a "blueprint" on which further regulatory actions could be built. The areas addressed include definition of terms used in UAS operations, the criteria on which UA flights into the NAS are authorized, the communication requirements between the UA and air traffic control (ATC), lost communication-link procedures, pilot responsibilities, and certification requirement for pilots and observers.

The FAA is currently engaged in the process of rule making for UAs. This will begin with the small UAs. Industry-led advisory panels, led by the Small UAS Aviation Rulemaking Committee, have concluded their work; it is likely the first notice of proposed rule making that is applicable to small UAs will be published in late 2010–early 2011. By late 2011, early 2012, the FAA should have regulations and procedures in place for all UAs.

5.3 DEFINITION OF AIRSPACE

The procedures governing UA are currently driven by the type(s) of airspace in which they are operating. Airspace classifications are designated by letters, and the rules for operations in each of the different classes vary. The airspace classifications in the United States are Class A, Class B, Class C, Class D, Class E, and Class G.

Class A—Class airspace begins at 18,000 feet above mean sea level (MSL) and extends to 60,000 feet above MSL. The majority of aircraft flying in this airspace are jet powered and fly at high speed. Some of these jets are small. Their small size, coupled with their fast speed, makes them difficult to see and avoid, hence, all aircraft in Class A airspace must operate under IFR regardless of weather conditions, and are separated by ATC. This is the case even though whenever visual meteorological conditions exist, visual separation must still be used. This airspace is called Positive Control Airspace. All UA operations in Class A airspace require the UA pilot to be certified, current, and instrument rated. UA observers are not required.

Class B—Class B airspace exists at very busy terminal areas. They usually surround larger cities, such as New York, Los Angeles, and Chicago. Due to the congestion of aircraft in a relatively small area, all aircraft must receive a clearance from ATC prior to entering Class B airspace, and all aircraft are separated by ATC, regardless of the weather conditions. UA operations in Class B airspace are prohibited.

Class C—Class C airspace exists around busy airports but not as busy as Class B. Aircraft must communicate with ATC prior to entering Class C airspace and comply with ATC instructions during operations in Class C airspace, regardless of weather conditions. All aircraft in Class C airspace must be equipped with a radar beacon transponder.

Class D—Class D airspace exists at airports with operating control towers. It generally begins at the surface of the airport and extends up to 2500 feet above the ground. Aircraft must communicate with ATC prior to entering Class D airspace, and comply with ATC instructions during operations in Class D airspace, regardless of weather conditions.

Class E—Class E airspace is referred to as controlled airspace. Aircraft may be flown in visual weather conditions (called visual flight rules, or VFR) without contact with ATC. If flight in IFR flight is anticipated, an ATC clearance must be obtained prior to entering instrument weather.

Class G—Class G airspace is considered uncontrolled airspace. ATC does not provide services to aircraft operating in Class G airspace. Hence, aircraft may fly in Class G airspace in any weather conditions with clearance or communication from ATC.

Certain sections of airspace must be avoided by all aircraft unless authorization to operate in them is obtained from the controlling agency. These special use airspace (SUA) areas are usually operated by the Department of Defense. Some types of SUA are prohibited areas, restricted areas, temporary flight restricted areas, and—in international airspace—warning areas. If permission is received from the controlling agency, UAs may be operated in SUA without a certificate of authorization or a special airworthiness/experimental certificate.

5.4 PUBLIC OPERATORS: THE CERTIFICATE OF AUTHORIZATION (COA)

The FAA has two categories of users of UAs. One is the public operator, and the other is the civil operator. Different procedures apply to UA flights apply to each of these. In this section, we will discuss public operators.

Who are public operators? Essentially, a public operator is any governmental institution. Examples include police, military, Department of Homeland Security, and state institutions of higher education. These operators are required to obtain a COA for every UA flight. The only exception to this are flights that are conducted wholly within restricted, prohibited, or warning areas with approval of the governing agency for that SUA. COAs are typically valid for one year as long as the UA operation is conducted within the constraints listed on the COA. COA application forms are submitted to the FAA, processed, and typically take at least 60 days for approval. However, they should be submitted at least 60 days prior to the proposed flight activity. Typically, submissions are done using FAA Form 7711-2. The COA must be received prior to the UA flight operation.

5.5 CIVIL OPERATORS: THE SPECIAL AIRWORTHINESS/ EXPERIMENTAL CERTIFICATE

Civil operators must have a special airworthiness (SAW) experimental certificate for a particular UA flight prior to a particular UA flight operation. Civil operators include all operators not under the category of public operators. Examples include individual citizens, private companies and organizations, and private educational institutions. Again, the only exception to this are flights that are conducted wholly within restricted, prohibited or warning areas with approval of the governing agency for that SUA. SAWs are typically valid for one year. SAW application forms are submitted to the FAA Airworthiness Division, in compliance with FAR 21.191 and may take several months for approval. The FAA requires the operator to have a continuing airworthiness program for all aircraft along with a maintenance training program.

5.6 FLIGHT OPERATIONS

Flight operations vary significantly, depending upon the size of the UA and the mission in which it is tasked. A UA the size of a Northrop Grumman Global Hawk (Figure 5.1) requires a runway; whereas an AeroVironment Dragon Eye (Figure 5.2) can be hand launched. Other UAs, such as the Boeing Insitu ScanEagle (Figure 5.3), require a catapult to launch for takeoff.

UAs that require a runway usually operate from an airport, and the collision avoidance issues departing from an airport are much greater than those launched from either a ramp or hand launched, because these may be done off-airport. Hence, the requirements for their COA or SAW will usually be much more extensive.

The missions flown by UAs vary widely. Small UA, such as the Dragon Eye, would typically fly missions no more than one hour in length. It weighs 5 pounds and is operated by an electric motor. It is controlled by use of a small onboard camera, which displays images on the pilot's laptop screen. It is flown within visual contact of the pilot. At the conclusion of the flight, the Dragon Eye can land in the grass near the pilot. Storage of the UA between flights is quite simple; the wings come off and it can be transported in a backpack.

The ScanEagle will usually fly more sophisticated missions. It has an autopilot, which allows autonomous flight. It can also be flown manually using a computer. The ScanEagle has the capability for "over-the-horizon" flight and can be flown hundreds of miles from the pilot using satellite relay communications. It is powered by a small, 2 horsepower gasoline engine, and has a ceiling of 20,000 feet and an endurance of 24 hours. The ScanEagle is usually operated from a small, portable ground control station (GCS). The GCS contains the computer display screens and UA controls, and is used during all flight operations. It has GPS, an array of different visual and electrical sensor suites, a transponder, and can be

FIGURE 5.1 Northrop Grumman Global Hawk.

FIGURE 5.2 AeroVironment Dragon Eye®.

FIGURE 5.3 Boeing ScanEagle®.

equipped with an automatic dependent surveillance broadcast (ADS-B) system, which will allow the pilot to see the ScanEagle wherever it flies by displaying its GPS-downlinked position on a chart in the GCS. The ScanEagle is recovered by flying it into a rope extending from a pole. A hook on the wing snags the rope (Figure 5.4).

FIGURE 5.4 Insitu ScanEagle recovery.

One of the largest UAs in use today is the Global Hawk. It has a wingspan of 130 feet, is powered by a Rolls Royce turbofan jet engine, can fly over 60,000 feet, cruise at 310 knots, and has an endurance of 36 hours. Equipped with an array of satellite and electro-optical sensors, the Global Hawk can carry a payload of 3000 pounds, and its 12,000-mile range allows it to fly over any area of the world as well as in Class A airspace. It requires a runway for takeoff and landing. Being nearly the size of a Boeing 737, the Global Hawk is normally stored in a hangar between flights.

The biggest challenge faced by UAS operators is the restrictions placed on the flight by the applicable COA or SAW certificate. Prior to any planned flight operations, the operator should refer to Section 8.2.14 of Interim Operational Approval Guidance 08-01. The UA should be equipped with the same avionics required for a manned aircraft in a particular class of airspace.

All flights below Class A airspace must be in VFR conditions, during daylight hours (exceptions may be approved to allow night flight), and usually require an observer, either standing on the ground watching the UA or in a chase aircraft. The purpose of the observer is to provide collision avoidance between the UA and manned aircraft.

When flying over oceanic areas, the same requirements apply when operating in airspace controlled by the FAA, even though the UA may be operating outside the borders of the United States. These areas are referred to as flight information regions (FIR).

5.7 PERSONNEL QUALIFICATIONS

As in UA flight operations, personnel qualification and requirements to operate a UAS are stated in FAA Interim Operational Approval Guidance 08-01 (see Table 5.1). Three job descriptions are mentioned: pilots, observers, and maintenance personnel.

Pilots—UA operations under IFR, and UA operations in Class A, C, D, and E airspace require the pilot be certificated in manned aircraft. Operation in IFR and Class A airspace also require an instrument rating. Additional operations that require a pilot certificate include nighttime operations, operations at joint civil–military or public-use airfields, and operations conducted

TABLE 5.1
UAS Personnel and Equipment Requirements in the National Airspace System

Airspace Class	UA Personnel Requirements	UA Equipment Requirements
A	*Pilot*—Pilot certificate with instrument rating and current second-class medical *Observer*—Not required	Compliance with FAR 91.135
B	UA operations not authorized	UA operations not authorized
C	*Pilot*—Pilot certificate and current second-class medical *Observer*—Required and current second-class medical	Transponder mode C or S
D	*Pilot*—Pilot certificate and current second-class medical *Observer*—Required and current second-class medical	Compliance with FAR 91.129
E	*Pilot*—Pilot certificate and current second-class medical *Observer*—Required and current second-class medical	Compliance with FAR 91.127
G	*Pilot*—No pilot certificate required if UA is flown under conditions mentioned in FAA Policy 08-01 Section 9.1.1.1; must have current second-class medical *Observer*—Required and current second-class medical	

beyond line-of-sight of the pilot. If a manned certificate is required, the pilot must be current as prescribed by Federal Aviation Regulation §61.57. The pilot in command, commonly referred to as the PIC, is the person ultimately responsible for the safe operation of the UA. The PIC must possess a current, second-class medical certificate. ALL UAS pilots must receive training in the UA being operated, including normal, abnormal, and emergency procedures. They must demonstrate proficiency and pass applicable testing on the UA being operated. The PIC may not simultaneously perform the duties of an observer during UA flight operations.

Observers—Observers are not required to possess a pilot certificate. However, they must possess a current second-class medical certificate. They must have an understanding of federal aviation regulations applicable to the airspace through which the UA is flying. In addition, they must have received training in federal aviation regulations concerning collision avoidance, in-flight right-of-way procedures, basic VFR weather minimums, and ATC phraseology.

Maintenance personnel—Currently, no certification is required for UA maintenance personnel. There are no medical requirements.

As this chapter closes, it is important to understand that the operation of UAS is evolving continuously along with the rapid advancement of technology. As proposed regulations become law and the industry growth levels off in coming years, we will begin to see some stabilization in the pace of change for UAS operations in the future. As such, it will be important to stay abreast of industry trade publications such as the Association for Unmanned Vehicle Systems International (AUVSI) *Unmanned Systems*, a monthly publication.

DISCUSSION QUESTIONS

5.1 Why is governmental regulation important for the UAS industry?
5.2 Should UAV pilots be rated in manned aircraft?
5.3 Should observers be required for UAV operations?
5.4 What are some operations in which UAVs can be effectively used in nonmilitary operations?

6 Unmanned Aircraft Systems for Geospatial Data

Caitriana M. Steele and Lisa Jo Elliott

CONTENTS

6.1 INTRODUCTION

Increasingly, consumer organizations, businesses, and academic researchers are using unmanned aircraft systems (UASs) to gather geospatial, environmental data on natural and man-made phenomena. These data may be either remotely sensed or measured directly (e.g., sampling of atmospheric constituents). The term *geospatial data* refers to any data that are referenced spatially, with a coordinate system, projection information, and datum. The low cost and easy deployment of UAS relative to manned aircraft and satellites means that UAS can respond rapidly to collect geospatial data on an expected or unexpected event or during a disaster (Ambrosia et al., 2007). Further, UASs may be used to monitor gradual changes such as fruit ripening for harvest (Berni et al., 2008). This chapter starts with a synopsis of the reasons for the growth in the popularity of remote sensing on UAS platforms, the types of sensors in use, and the image processing requirements. The chapter finishes with a review of some civilian applications for geospatial data acquisition.

6.1.1 Unmanned Aircraft Systems for Remote Sensing

Much of the growth in the use of UASs for geospatial data acquisition has been in the field of remote sensing. Remote sensing is the observation of the Earth's surface using instruments that measure reflected or emitted electromagnetic radiation; these instruments produce data that is usually represented in an image format (Campbell, 2007). A variety of imaging sensors have been used onboard UASs (see Section 6.1.2). Data from these sensors must be georectified in order to be used in a geospatial context (see Sections 6.1.3 and 6.1.4). Without georectification, imagery cannot be further processed or analyzed using geographic information systems (GIS). (GIS is a collective term used to describe the system of hardware, software, and standard operating procedures where geospatial data are organized, stored, analyzed, mapped, and displayed.)

Small- to medium-sized UASs are easily deployed so they can be very useful for gathering remote-sensing data at short notice. Many UAS platforms require little or no runway for takeoff or landing. The helicopter-style UAS requires no runway at all, but even fixed-wing systems can be launched in limited space or inhospitable areas. Fixed-wing UASs may be configured with either a vertical takeoff and landing (VTOL) system or a conventional takeoff and landing (CTOL). Similar to helicopters, VTOL systems lift directly over the launch point, but instead of an overhead rotor, VTOL UASs are equipped with "ducted fanned" apparatus. Similar to an air ventilation fan (i.e., a bathroom fan), this apparatus consists of a propeller mounted inside a vertically oriented cylindrical tube. Another advantage of UASs configured with VTOL is their maneuverability and their ability to "hover and stare" making these vehicles particularly well suited to urban and complex environments (Newman, 2006). In hover-and-stare mode, the UAS may transmit live data feed over a single object or event, directly to a ground station in real time (Newman, 2006).

Ease of deployment facilitates frequent UAS launches, which enables frequent data acquisition. Sensors on board a UAS can provide data more frequently than sensors on board most piloted aircraft or satellites (Nebiker et al., 2008; Puri et al., 2007). Frequent launches convey an important advantage: sensor data is closer to real time than data from manned flights or space-based satellites (Puri et al., 2007). Sensor data that are many hours old are of limited use in time-sensitive activities such as firefighting or rescue. The long duration flight ability of some UAS platforms have the potential to collect ongoing and near real-time data for agriculture (Furfaro et al., 2007; Herwirtz et al., 2002), traffic monitoring (Heintz, 2001; Puri et al., 2007) and disaster relief (Gerla and Yi, 2004).

The UAS advantages of maneuverability, ease of deployment, frequent data acquisition and fine spatial resolution converge with the advantage of safety in hazardous environments such as the Arctic (Inoue et al., 2008), overactive fires (Ambrosia et al., 2007), and during severe storms (Eheim et al., 2002). The Oliktok Point Arctic Research Facility (OPARF) in North Slope County, Alaska, allows researchers to track the melting of the sea ice over the Arctic with UAS. A joint agreement between Sandia National Labs, Department of Energy (DoE), and the Federal Aviation Administration (FAA), has created a warning area (similar to restricted airspace). In this area, unmanned aircraft flights and other research activities may be conducted over the Arctic Ocean (S. B. Hottman, personal communication, July 6, 2010).

6.1.2 Sensors

Sensors used onboard UAS include simple sensors such as (a) consumer-grade digital still and video cameras that measure reflected radiation in just three wavelengths: blue, green, and red; (b) multispectral frame cameras and line scanners that can sample reflected radiation in near infrared (NIR) and shortwave infrared (SWIR) wavelengths; and (c) sensors that measure emitted radiation in thermal infrared (TIR) wavelengths. The mission objective is an important criterion when choosing which type of imaging sensor is appropriate for a particular application. An example of this is the remote sensing and mapping of vegetation. For this application, it is desirable to have a sensor that can capture the unique spectral response of green vegetation between 600 and 900 nm (Hunt et al., 2010). To measure this type of data, the sensor must be able to obtain data in red and NIR wavelengths. Likewise, for fire detection and monitoring missions, thermal imagery combined with shortwave infrared and visible data is advantageous (Ambrosia, 2001; Ambrosia, Wegener, Brass et al., 2003; Ambrosia, Wegener, Sullivan et al., 2003).

In addition to mission considerations, payload capacity can impose restrictions on the choice of sensor used on board a UAS; the total weight of the payload should not exceed 20% to 30% of the weight of the system (Nebiker et al., 2008). Smaller UASs, such as the MLB Bat 3 or the Vector P™, have capacities up to 5 lbs and 10 lbs, respectively. At the other extreme, the Altus® II can carry up to 330 lbs in a nose payload compartment and the Altair® can carry up to 700 lbs. Table 6.1 lists sensors that have been used on board UASs for a variety of missions. With a low payload capacity, small UASs are constrained to consumer-grade digital still cameras or video cameras. If a mission requires a multispectral sensor, the payload capacity may exceed the capability of a small UAS. If a mission planner has access to a large UAS, sensors options are more generous. Many sensors available to large UASs are equivalent to those carried on manned aircraft (see, for example, Herwitz et al., 2002).

However, limited payload capacity has also been the inspiration for innovative adaptations of existing sensor technology. Hunt et al. (2010) explored adapting a single lens reflex (SLR) digital camera into an NIR-green-blue sensor for agricultural applications. Charge-coupled device (CCD) and complementary metal oxide semiconductor (CMOS) sensors are sensitive to radiation in NIR and visible wavelengths. To avoid NIR contamination, most cameras are fitted with an internal IR

TABLE 6.1
Examples of Sensors Used Onboard UAS Platforms

Sensor (wavelengths)	Platform(s)	Payload	Purpose	Reference
Canon Powershot S45, compact digital camera	MARVIN helicopter	2.2 lbs	Fire detection and monitoring	Ollero et al., 2006
Sony SmartCam, smart camera				
Canon EOS 20D, [single lens reflex digital camera]	weControl helicopter	2.2 lbs	Precision agriculture	Nebiker et al., 2008
Canon Digital Elph SD550, compact digital camera	MLB Bat 3	2.5–5 lbs	Rangeland monitoring	Laliberte et al., 2008
Fujifilm FinePix S3 Pro, single lens reflex digital camera	IntelliTech Microsystems Vector-P	10 lbs	Precision agriculture	Hunt et al., 2010
Tetracam MCA-6, six-channel multispectral camera	Benzin Acrobatic helicopter	~10 lbs	Precision agriculture	Berni et al., 2008, 2009
FLIR Systems Thermovision A40M (7.5–13 μm), thermal infrared digital video, machine vision camera				
Olympus C3030, compact digital camera	Aerosonde	12 lbs	Mapping melt ponds in sea ice	Inoue et al., 2008
Hassalblad 555ELD, large format film camera with Kodak Professional DCS Pro Back CCD array	Pathfinder-Plus	Up to 150 lbs	Precision agriculture	Herwitz et al., 2002
DuncanTech MS3100, purpose-built airborne digital multispectral imager				
Daedelus Airborne Large Format Imager, purpose-built airborne digital multispectral imager				
Airborne Infrared Disaster Assessment System (AIRDAS), purpose-built airborne digital multispectral-infrared scanner	ALTUS II	Up to 330 lbs		Ambrosia, Wegener, Brass et al., 2003; Ambrosia, Wegener, Sullivan et al., 2003

Note: Sensors and UASs are listed according to payload capacity.

cut filter (IIRCF). A camera without an IIRCF can be adapted to measure in NIR wavelengths. To do this, Hunt et al. (2010) used an interference filter to block red wavelengths and produce NIR-green-blue imagery.

Another option for meeting the payload constraints of small UASs are small, purpose-built multispectral cameras. For example, the company Tetracam manufactures the Agricultural Digital Camera (ADC), which is specifically designed to record radiation in red, green, and NIR wavelengths. Tetracam also produces a more advanced instrument, the Multiple Camera Array (MCA), which can include up to six spectral channels (Berni et al., 2008).

6.1.3 Real-Time Data Transmission

For some applications, especially those that are not time sensitive, imagery can be stored locally on a memory card in the camera. However, there are many situations that require real-time transmission or real-time processing of image data. If storage is not possible or real-time processing is required, sensor data may be sent back to the UAS operator or sent through an information processing unit. The data may be used for real-time navigation, to track areas of interest, or to collect specific information.

6.1.4 Georectification and Mosaicing of Still Imagery

Remotely sensed imagery requires a spatial reference before it can be used in a GIS. This can be as simple as pairing the center point of an image (with known scale) with spatial coordinates. For example, a UAS-borne sensor could be used to capture individual, noncontinuous frames as a means of sampling a target site (see, for example, Inoue et al., 2008). In this case, pairing of imagery with coordinates may be sufficient without further image rectification. Although this process should be straightforward, location errors may arise from improper synchronization between the onboard GPS and the sensor and insufficient GPS occupation time (Hruska et al., 2005). It is important to acknowledge that simple pairing of imagery with coordinates does not reconstruct the orientation of the sensor with respect to the target. Therefore, distances measured in the image will be relatively less precise when compared to distances measured on the ground. Notably, if there is no correction for variations in the altitude, attitude, and speed of an airborne sensor platform, then geometric distortions in the imagery may be too pronounced for mapping work.

Often, GIS applications require more precise georectification, where each pixel in the image is tied as closely as possible to its relative position on the ground. This is particularly true for precision agriculture and vegetation monitoring studies (Rango et al., 2006). There are photogrammetric procedures for processing imagery from airborne area (frame) and line sensors and registering this imagery to a coordinate system. However, these procedures do not always translate readily to UAS imagery (Laliberte et al., 2008). For example, conventional techniques such as aerial triangulation, image-to-image, or image-to-map registration require points that have known coordinates on the ground and that are detectable in the imagery (ground control points, GCPs). Because conventional techniques for image registration do not account for systematic distortions in the data (such as those caused by variations in the position of the platform relative to the target), they require many GCPs. This is particularly true of the large-scale imagery acquired by low-altitude, short-endurance (LASE) UASs. The cost of acquiring so many GCPs may be so high that it outweighs the advantages of acquiring the UAS imagery in the first place (Hruska et al., 2005). The cost associated with the collection of GCPs is a function of location. For example, detecting GCPs for fine-resolution imagery over an urbanized area is relatively straightforward because unique man-made features can usually be identified in both the imagery and on the ground. In an undeveloped area such as rangelands, an analyst would struggle to identify GCPs on the ground and in the imagery. Further, GCPs may be impossible to obtain either because the target of interest is relatively featureless or is inaccessible (see, for example, Inoue et al., 2008).

Imagery can be directly georectified without the use of GCPs using a photogrammetric approach (Hruska et al., 2005). To do this, the interior orientation (IO) parameters of the camera must be known (radial lens distortion, principal point offset, and focal length) (Berni et al., 2008, 2009; Laliberte, 2008). Metric cameras are supplied with these data but consumer-grade digital cameras are nonmetric and are not supplied with IO parameters. Fryer (1996) and Fraser (1997) explain the procedures necessary for characterizing the IO parameters of nonmetric cameras. Added to the camera IO parameters, the photogrammetric approach requires that exterior orientation (EO) parameters are measured simultaneously with image exposure. These parameters describe the position in space and perspective orientation of the sensor relative to the map coordinate system (Hruska et al., 2005). EO parameters include (a) aircraft roll, pitch, and yaw recorded by an onboard inertial measurement unit (IMU); and (b) aircraft elevation and position (latitude and longitude) as recorded by the GPS. For frame imagery, once the IO and EO parameters are known, they can be incorporated into a model for transforming the image from relative file coordinates to absolute map coordinates.

When topographic data such as elevation data are also used in georectification, the process is called orthorectification. Orthorectification is often used for the geometric correction of imagery from satellite and airborne sensors because it can yield a highly accurate representation of the Earth's surface (e.g., the U.S. Geological Survey [USGS] digital orthophoto quarter quads). The challenge for using digital elevation data to orthorectify very fine spatial resolution imagery from a UAS is the level of spatial detail of the digital elevation models (DEMs). These DEMS were originally created from contour maps and may contain artifacts from the process of conversion from line to raster data. In some areas, the artifacts are very pronounced and may introduce error into the geometric correction.

Georectification of line imagery is more complicated than for frame imagery. Line sensors usually contain a single line of detectors, which are repeatedly scanning. The forward motion of the platform means that successive lines are acquired to build an image. For these sensors, EO parameters must be modeled as a function of time.

The results of image georectification will also depend in part on the proximity of the sensor to the target. For example, LASE UASs are often operated at low altitudes above the target of interest. Linder (2009) summarizes the problem thus: smaller distances between the target and the camera, combined with a wide lens angle result in greater angles describing the central perspective and greater image distortion. Furthermore, individual frames have a small footprint compared to the exterior orientation parameters of the sensor. Image distortion can be exacerbated by gusts of wind and atmospheric turbulence to which small UASs are susceptible.

The small footprint of imagery obtained by a small UAS presents another image-processing challenge. To obtain a synoptic view of an area that is covered by individual frames, it is necessary to mosaic the imagery. This is an involved process that stitches multiple individual images together to create a single, large image covering the area of interest.

6.2 APPLICATIONS

6.2.1 Environmental Monitoring and Management

6.2.1.1 Precision Agriculture

Precision agriculture (PA) is a system that seeks to maximize long-term production and efficiency while optimizing resource use and sustainability. "Within field" spatial variability of soils and crop factors has long been recognized and attended to by farmers. As field size has grown and agricultural practices intensified, it has become increasingly difficult to address this variability without increasing reliance on technology (Stafford, 2000). PA requires spatial-distributed data on multiple soil, crop, and environmental variables. Moreover, these data must be collected and processed frequently enough so that the farmer has time to respond to key physiological developments in the crop or changes (e.g., pest damage, disease detection, nutrient or water stress, harvest readiness). Key technological developments in GPS, GIS, and remote sensing have helped to revolutionize PA. Of these key developments, the introduction of UAS as remote-sensing platforms has proved so successful that PA is one of the fastest growing civilian UAS applications.

For example, the National Aeronautics and Space Administration (NASA) tested the solar-powered Pathfinder Plus UAV as a long-duration platform for collecting imagery over a commercial coffee plantation (Furfaro et al., 2007; Herwirtz et al., 2002). One aim of this work was to detect the development of coffee bean ripeness during the 2002 harvest season. A long duration UAV (with wireless network connectivity to a ground station) provided near real-time monitoring of ripening. This enabled the farmer to identify the optimum time for harvest. The UAV used a DuncanTech MS3100 multispectral camera to acquire repeat imagery in green (550 nm), red (660 nm), and near-infrared (790 nm) wavelengths and with a spatial resolution of 1 m (Herwirtz et al., 2002). This spatial resolution is too coarse to resolve individual cherries. To detect ripeness, the contribution of fruit on the canopy surface to photon scattering and absorption was modeled with a modified leaf/canopy radiative transfer model (Furfaro et al., 2007). The model was then inverted using a neural network algorithm to estimate percentages of green, yellow, and brown cherries. Model estimates of ripeness were well correlated with yield data ($r = 0.78$) and even outperformed a ground-based assessment of harvest readiness.

An important aspect of remote sensing for precision agriculture is the use of radiometrically calibrated, remotely sensed data for deriving estimates of crop physiological properties. The study by Berni et al. (2008, 2009) provides a good example of the use of calibrated reflectance data for estimating crop leaf area index (LAI), canopy chlorophyll content, and crop water stress. These groups conducted intensive remote-sensing campaigns over cornfields, peach orchards, and olive orchards. They used a UAS airframe based on a model helicopter (Benzin Acrobatic, Germany) with a payload of a six-band multispectral frame camera (MCA-6, Tetracam Inc.) and a thermal frame sensor (Thermovision A40M, FLIR Systems).

Multispectral sensors measure the reflectance of vegetation in multiple discrete bands. Vegetation indices (VIs) use linear combinations of these bands (e.g., difference, ratio, or sum) to transform multiple spectral variables to a single spectral variable, which may then be related to vegetation canopy properties. Berni et al. (2008) investigated the relationships of several VIs with canopy temperature, LAI, and chlorophyll content. They found strong empirical relationships between (a) normalized difference vegetation index (NDVI) and olive LAI ($R^2 = 0.88$), (b) the physiological reflectance index (PRI) and corn canopy temperature ($R^2 = 0.69$), and (c) a variant of the transformed chlorophyll absorption in reflectance index (TCARI) and chlorophyll content in olive and peach canopies ($R^2 = 0.89$).

Surface temperature is important for detecting crop water stress (using the crop water stress index, CWSI) and can also be used to estimate canopy conductance. After calibration, the thermal imager was able to successfully estimate absolute surface temperatures over olive orchards (Berni et al., 2008, 2009). A particular advantage of the UAS thermal imagery was that its fine spatial resolution (40 cm) allowed for tree canopies to be distinguished from the soil background (Berni et al., 2009). This is not possible with the coarser imagery from satellite-borne sensors (e.g., Terra ASTER Thermal data have a spatial resolution of 90 meters).

6.2.1.2 Rangeland

Remote sensing by UASs has been very useful to rangeland management. About 50% of the Earth's land surface can be classified as rangeland. Globally, land management agencies share the same challenge: how to best monitor and manage rangeland resources over vast areas. For example, the Bureau of Land Management (BLM) manages around 258 million acres of land, mostly in the western United States. With a budget of $1 billion, this means that there is only $3.87 available per acre per year (Matthews, 2008). Remote sensing has been touted as a potential tool for assisting in monitoring and assessment of rangeland health. Remote sensing approaches can provide complementary information for managers and decision makers but as landscape complexity increase, the usefulness of remote sensing approaches decreases. Primarily, this is again a problem of sensor spatial resolution. Fine spatial resolution is a key requirement for the remote sensing of vegetation communities in arid and semiarid rangelands where concepts of rangeland health are tied closely to the distribution and connectivity of patches of vegetation, notably perennial grasses and woody shrubs (Bestelmeyer, 2006). Fine spatial resolution can help with identification and mapping of invasive plant species and localized surface disturbances (Matthews, 2008). As a tool for rangeland management, imagery from UASs hold great promise for extrapolating point-in-space ground surveys conducted by the range condition expert to a much wider area. UAS imagery could be used to scale between ground surveys to the regional view provided by satellite-borne sensors (Matthews, 2008).

In an example of a rangeland management application, Laliberte et al. (2008) have investigated how imagery from a small digital camera mounted on an MLB Bat 3 can be used to classify vegetation over rangelands in southern New Mexico. This group mosaiced multiple images together to form a synoptic view of each study site. The resulting image was "true color" with a spatial resolution of 5 cm. This imagery was classified to shrub, grass, forb (i.e., nongrass herbal vegetation), and bare ground using object-based image classification software. The level of detail in the resulting image classifications has great potential for incorporation into the current methods of range assessment used by the BLM.

The potential of UASs for rangeland management extends beyond land resources. Real-time or still imagery from UAS can be used in wildlife inventory (Matthews, 2008). Aerial platforms have been widely used for surveying animals, birds, nests, or food caches (Jones et al., 2006).

6.2.1.3 Ocean and Coastal Research

In 2005, the National Oceanographic and Aeronautics Administration (NOAA) conducted three successful test flights with the Altair UAS. One purpose of these flights was to examine the applicability of the imaging payload (ocean color imager, digital camera system, electro-optical infrared sensor) for coastal mapping, ecosystem monitoring, and surveillance of commercial and recreational activities in coastal waters and marine sanctuaries (Fahey et al., 2006). The NOAA test flights had multiple objectives including (a) remote sensing of ocean color (important for detecting chlorophyll-a suspended in ocean surface layers); (b) mapping Anacapa Island and the coastal areas of two Channel Islands using a digital camera system and electro-optical infrared sensor; (c) measuring atmospheric profiles of temperature and water vapor (for detecting atmospheric rivers) using a passive microwave vertical sounder; and (d) measuring the atmospheric concentrations of halogenated gases using a gas chromatography–ozone photometer.

The Arctic Ocean is one of the most inaccessible and hazardous environments in which to conduct UAS remote sensing. However, UASs make a vital contribution to remote sensing these extreme environments. Several problems are associated with satellite-based remote sensing of these areas. Specifically, cloud cover creates an impenetrable blanket for optical sensors. In response to this problem, satellite microwave sensors have been used to estimate sea ice extent. However, the problem with data from microwave sensors such as the advanced microwave scanning radiometer for the Earth Observing System (AMSRE) or the special sensor microwave imager (SSM/I) is that their coarse spatial resolution obscures fine-scale melt patterns and the formation of melt ponds. This can lead to underestimates of sea-ice concentration and obscures the evolution of melt ponds. Data on melt pond evolution is important for representing sea-ice albedo feedback (i.e., the ratio of light reflected) in climate models (Inoue et al., 2008).

To address the possible underestimates of sea-ice concentration and to obtain cloud-free imagery of melt ponds in sea ice in the Beaufort Sea, Inoue and colleagues (2008) mounted an Olympus C3030 digital still camera on an Aerosonde UAS and flew it at an altitude of 200 m. This configuration acquired imagery with a ground resolution of 8 cm. Rather than aim for a continuous mosaic of the imagery, the camera was triggered every 30 seconds to provide discrete, geolocated images of the study area. A simple classification of each image was achieved using threshold values of red, green, and blue values recorded (or interpolated) for each pixel. Using this straightforward sampling methodology and the simple image thresholding approach, Inoue et al. (2008) found that their UAS-derived measurements of sea ice and melt pond fraction compared well with findings from other studies. Their measurements, as well as the findings from others showed that, moving north from 72.5°, both sea ice and melt ponds increase in area. These UAS-derived measurements were also used to demonstrate how SSM/I data underestimate sea-ice concentrations.

6.2.1.4 Contaminant Spills and Pollution

6.4 Currently, UASs are rarely used for detecting and monitoring contaminant incidents such as oil spills. Therefore, much of the academic literature concerning the use of UAS for oil spills focuses on oil pipeline surveillance. Allen and Walsh (2008) extend upon this application, suggesting the potential of UAS for replacing or complementing manned aircraft monitoring to assist response to oil or hazardous substance spills in terrestrial and marine environments. Allen and Walsh (2008) detail how UAS can detect the initial spill. In a marine environment, the ease with which a small UAS can be deployed assists with frequent updating of oil migration. UASs can also be used in remediation in the application of aerial dispersants, helping determine shoreline cleanup requirements, and for wildlife rescue and rehabilitation.

In comparison more academic literature exists on the study of the atmospheric sampling of pollutants. NASA is one of the leading agencies involved in this activity. According to Cox et al. (2006), data collection efforts include gathering data on air pollution, radiation (shortwave atmospheric heating), cloud properties, active fire emissions, fire plume assessment, O_2 and CO_2 flux measurements, aerosols and gas contaminants, cloud systems, and contrails. In these various missions, NASA uses formation flying. Three UASs are required for in situ sampling of the inflow region, the outflow region, and the convective core (Cox et al., 2006). One UAS is required for high-altitude sensing. Depending on the contaminant and the range, the requirements include at least a 10,000-km range, an endurance of at least 24 hours, and heavy payload capacity.

6.2.2 Traffic Sensing

Currently, there are several methods used to track and monitor traffic in state-based Department of Transportation (DoT) organizations across the United States. DoTs implement video cameras mounted on towers, pavement embedded detectors, portable pneumatic tubes, and manned aircraft. Satellites had been considered for visual monitoring, but due to the transitory nature of satellite orbits and the coarse spatial resolution of satellite-borne sensors, it is difficult to implement a consistent monitoring pattern (Puri et al., 2007). Many DoTs are starting to explore unmanned aircraft to replace the real-time visual monitoring of traffic during high use periods. Ultimately, many DoTs would like an autonomous system such as the one described by Heintz (2001):

> As the operator told the unmanned helicopter to watch the red Ford [car] driving at high speed on the highway the helicopter did a sharp turn and increased its velocity to catch up with the speeding car, containing an escaped prisoner and his accomplices. As the distance decreased the operator got continuous updates of the actions of the fleeing vehicle. While the helicopter watched the car and tried to anticipate the escape route, the operator guided the police to set up a road block where the criminals could be caught and arrested. (p. 1)

This scenario describes the goal of many UAS traffic monitoring systems currently in development. The objective of many of these systems is to navigate and plan autonomously. Organizations would like for the UAS to locate, identify, monitor, and continuously track a specific vehicle; to identify vehicle trajectories and unusual driver behavior; and to monitor intersections and parking lots (Heintz, 2001). In response, WITAS (Wallenberg Laboratory for Information Technology and Autonomous Systems), at Linköping University in Sweden, has created a prototype that integrates many of these functions (Heintz, 2001; Puri et al., 2007). This prototype and others include several types of remote sensors and a combination of different types of cameras. A further challenge to traffic monitoring is tracking a specific vehicle as it moves alongside other vehicles in varying directions and at varying speeds. Bethke et al. (2007) propose a combination of sensors for a prototype ground moving target indication (GMTI) sensor that is suitable for this task.

Another challenge to implementing traffic monitoring on a UAS is the canyon effect. The canyon effect refers to the UAS's ability to follow roads through mazes of high-rise buildings. The high-rise buildings create canyons where communication and visual contact is easily lost. Several developers are working on autonomous agents to address this issue. Ng et al. (2005) are developing game theoretic optimal deformable zone with inertia, or GODZILA. GODZILA has an advanced algorithm for path planning and obstacle avoidance in unknown environments. Other developers have taken a prior-knowledge approach and customized the UAS to work in a specific area by hard coding the building locations.

UASs offer many advantages over the current traffic monitoring and transportation planning for DOT, emergency response, and law enforcement teams. UASs can move at higher speeds and are not restricted to specific routes or the ground as are conventional vehicles. They may fly in dangerous or inclement weather conditions. UASs can deploy rapidly with very little runway space and can be virtually undetectable.

6.2.3 Disaster Response

In civilian applications, UASs provide an indispensible means of gathering data about a situation on the ground. UASs are particularly adaptable to missions in one or all of the three D's: dirty, dangerous, and dull. As with an unmanned ground vehicle (UGV), UASs in disaster response missions require some additional features that will allow them to operate in extreme situations. Jinguo et al. (2006) describe these as features as survivability, durability, and adaptability.

UAS survivability in disaster missions relies on efficient and extensive communication systems. UAS search-and-rescue communications must consider three aspects: communication between the operators and the UAS, communication between the operators and the victims, and communication between the other rescue machines and their teams. As in human teams, communication determines the ability of the system to adapt to changing circumstances and maintain situation awareness in the rapidly changing disaster environment.

UAS durability includes the system's ability to survive unpredictable dropping debris, operate in an uncertain and changing environment, and cope with loss of signal (LoS) problems. To offset these issues, designers have proposed that rescue teams employ not one but several levels of UASs on a team as human rescuers compose a multilevel team (Gerla and Yi, 2004). Murphy et al. 2006 (p. 176) suggest a "5:2 human–robot ratio with three co-located humans in the roles of Operator, Mission Specialist and Flight Director." A suitable rescue system might incorporate medium-sized UASs or a HALE (high altitude, long endurance) UAS to carry equipment, provide an interim communication link, and to provide an overview of the area with possible exit routes and information on changing conditions. More information on communications relays can be found in Gerla and Yi (2004). A small UAS might have similar tasks directly above an area of concern. Then, a mini-UAS (MUAS) is tasked with gathering data about on the ground conditions at a particular site and searching for survivors of the disaster. When the MUAS experiences LoS problems, there are additional and mobile access points for an intermediary signal. The MUAS also relies on the small UAS and the medium UAS for information about changing conditions, structural changes and possibilities of flying debris (Jinguo et al., 2006; Teacy et al., 2009).

UAS adaptability includes the ability for the MUAS to be small enough to overcome unpredictable debris trajectories and narrow spaces but with the ability to sense changes in an unstructured, uncertain environment (Jinguo et al., 2006). UAS adaptability also includes the capability of recording physical information and the exploration of unknown conditions. Disaster rescue professionals operate in teams with each team responsible for a different level of the disaster. Often, disaster relief operations are separated into three stages of "the pre-disaster rescue operation, the on-disaster rescue and the post-disaster rescue operation" (Jinguo et al., 2006, p. 439). Jinguo and colleagues (2006) describe the stages according to the disaster/rescue professional teams' expectations. In the pre-disaster preparations, the teams coordinate evacuations and prepare materials. During the on-disaster rescue, the teams fight the disaster and mitigate damages. The post-disaster teams search for and rescue survivors. Often there is no clear boundary between the rapidly changing stages. At different stages of the disaster, different rescue teams are active. UAS rescue teams should mimic this arrangement. A HALE UAS could be deployed for pre-disaster coordination of evacuation and traffic monitoring while the operators establish mobile ad-hoc communication systems (Gerla and Yi, 2004). As the disaster unfolds, additional levels of UAS are deployed for data gathering, relief package delivery, and search operations.

6.2.3.1 Fires

Satellite remote sensing of active fires is limited by the spatial resolution of the imagery and by the temporal frequency with which imagery can be acquired (Alexis et al., 2009; Casbeer et al., 2006). For example, the U.S. Department of Agriculture Forest Service Active Fire Mapping Program uses thermal imagery from moderate resolution imaging spectrometer (MODIS) sensors on board the Terra and Aqua satellites (USDA–USFS, 2010). Data are acquired twice daily by each sensor at a spatial resolution of 1 km. These data provide useful indications of fire activity at regional and national scales but the spatial resolution is too coarse to provide precise fire-front position information. Further, the revisit interval of the satellite sensors is too infrequent for tracking the evolution of the fire and managing firefighting in real time.

Two different approaches have been tested for detecting and monitoring forest fires with a UAS. The first uses HALE UASs, capable of long duration missions of multiple fires (Ambrosia, Wegener, Brass, et al., 2003). These systems can provide imagery with finer spatial resolution and more frequently than satellite sensors (Casbeer et al., 2006). The second approach uses fleets of LASE UASs working cooperatively (Alexis et al., 2009; Casbeer et al., 2006; Merino et al., 2006).

Ambrosia et al. (2001), Ambrosia, Wegener, Brass, et al. (2003), and Merlin (2009) report on the collaboration between NASA–Ames, General Atomics Aeronautical Systems Inc., and various government research agencies in the First Response Experiment (FiRE) projects. These projects used the General Atomics ALTUS® II UAS, a civil variant of the Predator®, mounted with a NERA M4 Mobile World Communicator System and an AIRDAS thermal multispectral scanner. The AIRDAS scanner was used to capture images of relative fire intensity over controlled fires. The NERA telemetry system then transmitted AIRDAS data and navigation files via INMARSAT geostationary satellites to the ground control station. Once image data were received at the ground station, they were georectified using Terra-Mar's Data Acquisition Control System (DACS) software. In the second FiRE project (Ambrosia, 2003a) digital elevation data from the shuttle radar topography mission (SRTM) were used to orthorectify the image data and to create three-dimensional models of the fire. Both FiRE projects proved that AIRDAS imagery and navigation data could be transmitted to the ground using a satellite image data telemetry system, georectified and then disseminated to Web (and so to potential users) in near real time (Merlin, 2009).

The FiRE projects and the establishment in 2003 of the Wildfire Research and Applications Partnership (WRAP) project (a collaborative effort between NASA and the U.S. Forest Service) built the foundation for the Western States Fire Mission (Ambrosia et al., 2007). The Western States Fire Mission continued to test HALE UASs for wildfire monitoring using first an Altair UAS and then the

Ikhana, both civilian models of the Predator B®. Both aircraft carried the autonomous modular scanner (AMS), which is a thermal multispectral scanner similar to the Airborne InfRared Disaster Assessment System (AIRDAS) instrument. Notably, the Western States Fire Mission graduated suddenly from experimental to operational status in October 2006, when NASA was awarded an emergency certificate of authorization to fly its Altair UAS in national airspace over the Esperanza fire. Esperanza was an arson-set fire in southern California that devastated over 40,000 acres and caused five fatalities (Ambrosia et al., 2007). As proven in the earlier FiRE projects, data were transmitted in near real time to the incident command center via the UAS ground control station (for georectification) and the Internet for viewing on Google Earth. The Western States Fire Mission continued to operate the Ikhana UAS (with the autonomous modular scanner) through 2007 for monitoring of eight other uncontrolled wildfires including two fires in the San Bernardino National Forest, one on the Camp Pendleton Marine Base, four in San Diego County, and one in the Cleveland National Forest in Orange County. As with the first operational excursion over the Esperanza fire, the Western States Fire Mission used the Ikhana and their ground support facilities to successfully acquire, transmit, process, and distribute imagery in near real time to support fire incident decision making at ground zero. The Western States Fire Mission also uses the Ikhana to acquire postfire imagery for burned area mapping. Figure 6.1 shows imagery that was acquired by the AMS on board the Ikhana of (a) the Zaca fire burn front and (b) postfire burned area. The Zaca fire started on July 4, 2007. By August 31, it was estimated to have burned over 240,000 acres.

HALE systems are expensive to procure and operate. In response to these constraints, other organizations use fleets of multiple, low-cost LASE UAS for detecting and monitoring fires. Whereas the HALE systems provide an image overview of the

FIGURE 6.1 The Zaca fire. (From NASA image gallery: http://asapdata.arc.nasa.gov/ams/gallery/fires/0704200.zaca_mosaic_12-10-9.jpg.)

entire fire, LASE systems detect the perimeter of the fire and transmit these data to a base station as frequently as possible. To accomplish this, each LASE UAS in a fleet must receive sufficient information for autonomous path planning and adjustment (Casbeer et al., 2006). The application of LASE UAS fleets for wildfire monitoring is largely confined to experimental and theoretical exercises but holds great potential. Alexis et al. (2009) have used simulation studies to demonstrate how a homogenous fleet of quadcopters (UqHs) can operate cooperatively using an independent cooperative control algorithm that depends on multiple rendezvous locations.

Other researchers have tested heterogeneous fleets of UASs working for cooperative monitoring of controlled fires. For part of the COMETS project, cooperation between two helicopters and an airship was tested over an experimental fire (Merino et al., 2006; Ollero et al., 2006). One helicopter (Helivision-GRVC) carried a Raytheon 2000AS thermal microcamera (7–14 μm) and a Camtronics PC-420DPB video camera. The second helicopter, "Marvin," carried a Hamamatsu UV-Tron fire detector and a Canon Powershot S40 digital still camera (Merino et al., 2006). The airship carried two digital IEEE1394 cameras for acquisition of stereo pairs. This stereo photography could then be used to visualize the terrain in three dimensions. The helicopter UASs were given areas to patrol until one UAS detected the fire. Once one helicopter detected the fire, the other was sent to the fire location to confirm. After a fire has been confirmed, the fire monitoring starts (Merino et al., 2006).

Ollero et al. (2006) suggest that fleets of UASs could complement the use of a single large HALE UAS. A single large UAS has the advantage of range and duration, and so can cover a large area for initial fire detection. Once a fire is detected, fleets of small UASs can then be used as a rapid response to confirm the presence of a fire or to flag it as a false alarm. If a fire outbreak is confirmed, then the UAS fleet remains to monitor the propagation of that fire (Ollero et al., 2006).

6.2.3.2 Floods and Hurricanes

Hurricane Wilma in 2005 was the first known use of unmanned vehicles to assess damage and assist in recovery efforts. Since then, researchers have been able to identify the types of technology needed and test the systems in mock recovery efforts. Rescue professionals have been able to share their experiences with each other. In their articles about the Katrina rescue effort, Murphy et al. (2006) and Pratt et al. (2006) outline possible rescue implementations and probable challenges.

During Hurricane Katrina, which hit New Orleans in 2005, several levels of UASs were proposed for deployment. Leitl (2005) describes a proposal to use small UAS and medium UAS. First, the Evolution, a small UAS, was proposed to assess structure and flooding damage. The Silver Fox, a medium UAS, was proposed to search for survivors with an IR camera. Additional teams in Mississippi were led by the University of South Florida and proposed possible deployment of similar UASs to search for trapped flood victims.

Despite the claims of researchers, it should be noted that FAA approval to operate UASs in the NAS was not granted for several safety reasons. The requirement that manned and unmanned aircraft operations should be segregated could not be fulfilled. Proposals to operate UASs during the storm were unable to demonstrate alternative communication capabilities. Overall, air traffic control capabilities in the area were limited due to the storm.

However, Leitl (2005) states that hurricane relief UAS deployments may have limited success in locating victims. When searching for victims, the location and the state of the victim are unknown. Because victim detection depends on the capabilities of the onboard UAS sensors, if the onboard sensors do not include IR, it can be difficult to assess if an object is a human body. It can also be difficult to determine if a still body remains alive. In response to this challenge, Doherty and Rudol (2007) propose a combination of IR and electro-optical cameras to detect the presence of life in nonmoving human forms.

6.2.3.3 Tornadogenesis

Many applications for UAS geospatial data collection relate to remote sensing, but UASs are also ideal platforms for atmospheric sampling in hazardous environments. For example, UASs provide a vital opportunity for improving tornado warning capabilities. Observations of the thermodynamic profile of the atmospheric column between the ground and the base of the mesocyclone, and particularly of the phenomenon known as the rear-flank region of the supercell, would greatly advance our understanding on the genesis and the evolution of tornadoes (Elston and Frew, 2010). Efforts to use piloted aircraft to collect these data have exposed both pilot and aircraft to unacceptable risk (Eheim et al., 2002).

As part of the Verification of the Origins of Rotation in Tornadoes Experiment (VORTEX), a team from the University of Colorado has been exploring the use of UASs for research into tornadogenesis. On May 6, 2010, the team launched a Tempest UAS to intercept the rear flank of a supercell thunderstorm. It was mounted with a sonde to measure air pressure, temperature, and moisture. A telemetry system continuously transmitted these data to a ground control station (Nicholson, 2010).

The use of UAS in tornadogenesis research provides some unique challenges. The UAS exoskeleton must be able to withstand heavy rain, ¾-inch hail, and 10 g loads of vertical gusts in order to remain within the rear flank of the supercell and successfully transmit data for 30 to 60 minutes (Erheim et al., 2002). On top of these requirements, the airframe must be inexpensive, because it is expected that the UAS could be destroyed during its mission (Erheim et al., 2002).

Aside from the engineering challenges of the UAS itself, tornadogenesis researchers face some unique problems when operating a UAS in national airspace. All UAS operators must receive a certificate of authorization (COA) from the FAA. Fulfilling the conditions of a COA, a UAS operator must submit a UAS flight map to the FAA 48 to 72 hours prior to a mission. For tornado chasers, the problem with these procedures for applying and receiving permission to fly is that it is not possible to predict with any certainty where a tornado might form. Further, the FAA requires that visual contact with a UAS be maintained at all times during its flight. This was a requirement for the Tempest flights. The risk of operation a UAS in pretornadic conditions and the potential distance that may need to be covered during tornadogenesis, led to an innovative albeit "ironic" solution. The flight computer of the Tempest was instructed to follow a manned ground vehicle (Nicholson, 2010). This allowed the aircraft to be positioned quickly and exactly where needed.

6.3 CONCLUSION

In this chapter we discussed the increasing popularity of UAS for the acquisition of geospatial data, particularly through remote sensing. We provided a brief summary of sensors used on board a variety of UAS and for different applications. Some of the challenges with georectification of still imagery were addressed. We have reviewed several applications for UAS remote sensing ranging from remote sensing of the natural environment, the managed environment, and the human environment. We also included two examples of UAS for *in situ* recording of geospatial data (Elston and Frew, 2010; Fahey et al., 2006). When the time comes that protocols are defined for obtaining permission to fly in national airspace, we anticipate the growth in the use of UASs for geospatial data collection will be exponential.

DISCUSSION QUESTIONS

6.1 What are some of the benefits of UASs over manned aircraft or satellites for collecting geospatial data?
6.2 What are some of the concerns with using a UAS for disaster relief?
6.3 What are some of the challenges UASs must overcome in order to gather traffic data in a high population area?
6.4 Discuss other dirty, dangerous, and dull missions that might lend themselves to gathering geospatial data via UASs. What constraints or challenges might the UAS and the sensors need to overcome to operate in hazardous environments?
6.5 The chapter purposefully omitted a discussion of sensory data analysis. What are some ways that an analyst might address the challenges created by voluminous amounts of sensory data?

REFERENCES

Allen, J., and B. Walsh. 2008. Enhanced oil spill surveillance, detection and monitoring through the applied technology of unmanned air systems. Proceedings of the 2008 International Oil Spill Conference, Savannah, GA. http://iosc.org/papers/2008%20019.pdf (accessed June 12, 2010).

Alexis, K., G. Nikolakoupoulos, A. Tzes, and L. Dritsas. 2009. Coordination of helicopter UAVs for aerial forest-fire surveillance. In *Applications of Intelligent Control to Engineering System, Intelligent Systems, Control, And Automation: Science and Engineering*, vol. 39, ed. K. P. Valavanis, 169–193. Netherlands: Springer.

Ambrosia, V. G. 2001. Remotely piloted vehicles as fire imaging platforms: The future is here! http://geo.arc.nasa.gov/sge/UAVFiRE/completeddemos.html (accessed June 12, 2010).

Ambrosia, V. G., B. Cobleigh, C. Jennison, and S. Wegener. 2007. Recent experiences with operating UAS in the NAS. American Institute of Aeronautics and Astronautics 2007 Conference and Exhibit, Rohnert Park, CA.

Ambrosia, V. G., S. S. Wegener, J. A. Brass, and S. M. Schoenung. 2003. The UAV Western States Fire Mission: Concepts, plans and developmental advancements. AIAA 3rd "Unmanned Unlimited" Technical Conference, Workshop and Exhibit, Chicago, IL.

Ambrosia, V. G., S. S. Wegener, D. V. Sullivan, S. W. Buechel, S. E. Dunagan, J. A. Brass, J. Stoneburner, and S. M. Schoenung. 2003. Demonstrating UAV-acquired real-time thermal data over fires. *Photogrammetric Engineering and Remote Sensing* 69: 391–402.

Berni, J. A. J., P. J. Zarco-Tejada, G. Sepulcre-Canto, E. Fereres, and F. Villalobos. 2009. Mapping canopy conductance and CWSI in olive orchards using high resolution thermal remote sensing imagery. *Remote Sensing of Environment* 113: 2380–2388.

Berni, J. A. J., P. J. Zarco-Tejada, L. Suarez, and E. Fereres. 2008. Thermal and narrowband multispectral remote sensing for vegetation monitoring from an unmanned aerial vehicle. *IEEE Transactions on Geoscience and Remote Sensing* 47: 722–738.

Bestelmeyer, B. T. 2006. Threshold concepts and their use in rangeland management and restoration: the good, the bad, and the insidious. *Restoration Ecology* 14: 325–329.

Bethke, K. H., S. Baumgartner, and M. Gabele. 2007. Airborne road traffic monitoring with RADAR. 14th World Congress on Intelligent Transport Systems. http://elib.dlr.de/51746/01/ITS-Paper_2243_Bethke.pdf (accessed July 7, 2010).

Campbell, J. B. 2007. *Introduction to Remote Sensing*. New York: The Guilford Press.

Casbeer, D. W., D. B. Kingston, R. W. Beard, and T. W. McLain. 2006. Cooperative forest fire surveillance using a team of small unmanned air vehicles. *International Journal of Systems Science* 37: 351–360.

Cox, T. H., I. Somers, D. J. Fratello, C. J. Nagy, S. Schoenung, R. J. Shaw, M. Skoog, and R. Warner. 2006. Earth observations and the role of UAVs. http://www.nasa.gov/centers/dryden/pdf/175939main_Earth_Obs_UAV_Vol_1_v1.1_Final.pdf (accessed June 12, 2010).

Doherty, P., and P. Rudol. 2007. A UAV search and rescue scenario with human body detection and geolocalization. In *Proceedings of Australian Conference on Artificial Intelligence*, pp. 1–13.

Eheim, C., C. Dixon, B. M. Argrow, and S. Palo. 2002. TornadoChaser: A remotely-piloted UAV for in situ meteorological measurements. Presented at AIAA's 1st Technical Conference and Workshop on Unmanned Aerospace Vehicles, Systems, Technologies, and Operations, Portsmouth, VA. http://recuv.colorado.edu/~dixonc/dixonc/Publications/papers/Eheim-Dixon_TornadoChaser.pdf (accessed June 12, 2010).

Elston, J., and E. Frew, 2010. Unmanned aircraft guidance for penetration of pretornadic storms. *Journal of Guidance, Control and Dynamics* 33: 99–107.

Fahey, D. W., J. H. Churnside, J. W. Elkins, A. J. Gasiewski, K. H. Rosenlof, S. Summers, M. Aslaksen, et al. 2006. Altair Unmanned Aircraft System Achieves Demonstration Goals. *EOS Transactions AGU* 80: 197–201.

Fraser, C. S. 1997. Digital camera self-calibration. *ISPRS Journal of Photogrammetry and Remote Sensing* 52: 149–159.

Fryer, J. G. 1996. Camera calibration. In *Close Range Photogrammetry and Machine Vision,* ed. K. B. Atkinson and J. G. Fryer, 156–179. Caithness, Scotland: Whittles Publishing.

Furfaro, R., B. D. Ganapol, L. F. Johnson, and S. R. Herwitz. 2007. Neural network algorithm for coffee ripeness evaluation using airborne images. *Applied Engineering in Agriculture* 23: 379–387.

Gerla, M., and Y. Yi. 2004. Team communications among autonomous sensor swarms. *SIGMOD Record* 33: 20–25.

Heintz, F. 2001. Chronicle recognition in the WITAS UAV Project: A preliminary report. Paper presented at the Swedish AI Society Workshop, Skövde, Sweden.

Herwitz, S. R., L. F. Johnson, J. F. Arvesen, R. G. Higgins, J. G. Leung, and S. E. Dunagan. 2002. Precision agriculture as a commercial application for solar-powered unmanned aerial vehicles. 1st AIAA UAV Conference, Portsmouth VA.

Hruska, R. C., G. D. Lancaster, J. L. Harbour, and S. Cherry. 2005. Small UAV-acquired, high-resolution, georeferenced still imagery. Proceedings of AUVSI Unmanned Vehicle Systems North America, Baltimore, MD.

Hunt, E.R., W. D. Hively, S. J. Fujikawa, D. S. Linden, C. S. T. Daughtry, and G. W. McCarty. 2010. Acquisition of NIR-green-blue digital photographs from unmanned aircraft for crop monitoring. *Remote Sensing 2010* 2: 290–305.

Inoue, J., J. A. Curry, and J. A. Maslanik. 2008. Application of aerosondes to melt-pond observations over Arctic sea ice. *Journal of Atmospheric and Oceanic Technology* 25: 327–334.

Jinguo, L., W. Yuechao, L. Bin, and M. Shugen. 2006. Current research, key performances and future development of search and rescue robots. *The Journal of Mechanical Engineering* 42: 1–12.

Jones, G. P., L. G. Pearlstine, and H. F. Percival. 2006. An assessment of small unmanned aerial vehicles for wildlife research. *Wildlife Society Bulletin* 34: 750–758.

Laliberte, A., C. Winters, and A. Rango. 2008. A procedure for orthorectification of sub-decimeter resolution imagery obtained with an unmanned aerial vehicle (UAV) [abstract], Proceedings of the American Society for Photogrammetry and Remote Sensing Annual Conference, April 28–May 2, 2008, Portland, OR. http://www.asprs.org/publications/proceedings/portland08/0046.pdf (accessed June 12, 2010).

Leitl, E. 2005. Information technology issues during and after Katrina and usefulness of the Internet: How we mobilized and utilized digital communications systems. *Critical Care* 10: 110.

Linder, W. 2009. *Digital Photogrammetry: A Practical Course*. Berlin: Springer-Verlag.

Matthews, N. A. 2008. Aerial and close-range photogrammetric technology: Providing resource documentation, interpretation, and preservation. Technical Note 428. U.S. Department of the Interior, Bureau of Land Management, National Operations Center, Denver, Colorado.

Merino, L., F. Caballero, J. R. Martinez-de Dios, J. Ferruz, and A. Ollero. 2006. A cooperative perception system for multiple UAVs: Application to automatic detection of forest fires. *Journal of Field Robotics* 23: 165–184.

Merlin, P. W. 2009. *Ikhana. Unmanned Aircraft System, Western States Fire Missions.* Monographs in Aerospace History #44, NASA SP-2009-4544 http://www.aeronautics.nasa.gov/ebooks/downloads/ikhana_monograph.pdf (accessed June 12, 2010).

Murphy, R. R. 2006. Fixed- and rotary-wing UAVS at Hurricane Katrina. In IEEE International Conference on Robotics and Automation (video proceedings), Orlando, FL.

Murphy, R. R., C. Griffin, S. Stover, and K. Pratt. 2006. Use of micro air vehicles at hurricane Katrina. In IEEE Workshop on Safety Security Rescue Robots, Gaithersburg, MD.

Nebiker, S., A. Annen, M. Scherrer, and D. Oesch. 2008. A lightweight multispectral sensor for micro UAV opportunities for very high resolution airborne remote sensing. *The International Archives of the Photogrammetry, Remote Sensing and Spatial Information Sciences*, ISPRS Congress, Beijing, China, XXXVII. Part B1, 1193–1199. http://www.isprs.org/proceedings/XXXVII/congress/1_pdf/204.pdf (accessed June 12, 2010).

Newman, D. 2006. Micro-Air Vehicle (MAV) Demonstrated Backpackable Autonomous VTOL UAV Providing Hover and Stare RSTA to the Small Military Unit. 25th Army Science Conference, Grande Lakes, Orlando, FL. http://www.dtic.mil/cgi-bin/GetTRDoc?AD=ADA481068&Location=U2&doc=GetTRDoc.pdf (accessed June 12, 2010).

Ng, T. L., P. Krishnamurthy, F. Khorrami, and S. Fujikawa. 2005. Autonomous flight control and hardware-in-the-loop simulator for a small helicopter. Proceedings of IFAC WC, Prague, Czech Republic.

Nicholson, C. 2010. Droning it in: Storm-chasing unmanned aerial vehicle makes first foray into nascent twister. *Scientific American,* http://www.scientificamerican.com/article.cfm?id=droning-it-in-storm-chasing-twiter (accessed June 12, 2010).

Ollero, A., J. R. Martínez-de-Dios, and L. Merino. 2006. Unmanned aerial vehicles as tools for forest-fire fighting. V International Conference on Forest Fire Research, Coimbra, Portugal. http://grvc.us.es/publica/congresosint/documentos/2006VICFFR_AOLLERO.pdf (accessed June 12, 2010).

Ortiz, G.G., S. Lee, S. Monacos, M. Wright, and A. Biswas. 2003. Design and development of a robust ATP subsystem for the Altair UAV-to-Ground Lasercomm 2.5 Gbps demonstration. In *Free-Space Laser Communication Technologies XV*, ed. G. S. Mecherle, Proceedings of the SPIE 4975: 103–114. http://opticalcomm.jpl.nasa.gov/PAPERS/ATP/gospie03.pdf (accessed June 12, 2010).

Pratt, K., R. R. Murphy, S. Stover, and C. Griffin. 2006. Requirements for semi-autonomous flight in miniature UAVs for structural inspection. Proceedings of AUVSI Unmanned Systems North America, Orlando, FL.

Puri, A., K. Valavanis, and M. Kontitsis, 2007. Generating Traffic Statistical Profiles Using Unmanned Helicopter- Based Video Data. In *2007 IEEE International Conference on Robotics and Automation*, 870–876. Rome, Italy.

Rango, A., A. Laliberte, C. Steele, J. E. Herrick, B. Bestelmeyer, T. Schmugge, A. Roanhorse and V. Jenkins. 2006. Using unmanned aerial vehicles for rangelands: Current applications and future potentials. *Environmental Practice* 8: 159–168.

Stafford, J.V., 2000. Implementing Precision Agriculture in the 21st Century. *Journal of Agricultural Engineering Research* 76: 267–275.

Teacy, W. T. L., J. Nie, S. McClean, G. Parr, S. Hailes, S. Julier, N. Trigoni, and S. Cameron. 2009. Collaborative sensing by unmanned aerial vehicles. Proceedings of the 3rd International Workshop on Agent Technology for Sensor Networks, May 2009, Budapest, Hungary. http://web.mac.com/lteacy/ATSN-09/proceedings_files/suaave.pdf (accessed June 12, 2010).

USDA-USFS. 2010. Active Fire Mapping Program. http://activefiremaps.fs.fed.us/faq.php (accessed June 12, 2010).

7 Automation and Autonomy in Unmanned Aircraft Systems

Lisa Jo Elliott and Bryan Stewart

CONTENTS

7.1 AUTOMATION AND AUTONOMY

For decades, system designers and system operators have struggled with automation. System designers predict decreased workload, increased precision, and better system performance. System operators work with imperfect automation, system failures, and automation-induced accidents. Yet, we rely on automation to regulate the temperature of our house, make our coffee, back up our computers, and do the menial work of our daily lives. Somehow, the automation improves and the promises are fulfilled.

The struggle between system designers and system operators mirrors the struggle between the human factors (HF) researchers and the automation designers in engineering. Clearly, automation throughout the decades has improved as a result. In this chapter, we will give a brief overview of the HF research in automation and then provide an overview of current and future unmanned aircraft system (UAS) automation efforts.

Automation enables UASs to have the capabilities and procedures needed for a UAS to fly in manned airspace. Moray et al. (2000, p. 44) define automation as "any sensing, detection, information-processing, decision-making, or control action that could be performed by humans but is actually performed by machine." As this definition implies, automation can be implemented at various levels of a system. Automation is not all-or-nothing; rather it is an agent that interacts with the human operator (hereto referred to as "operator"). Regardless of the designer's intent, automation has behavior. This behavior interacts with the operator's behavior, the operator's mental model of the system, and the operator's trust of the system. As a result, automation itself changes an operator's training, task assignments, workload, situation awareness, trust, and even the operator's skill set.

As Woods (1996) states, automation is not "adding another team member." Automation changes the dynamics between the operator and the system. The automation is limited in its ability; it is not a full "team player." A system and its automation are deaf, cannot freely communicate, and have only the abilities that the system designer has deemed necessary. The unequal distribution of "team" responsibilities results in what Woods calls *automation surprises* (Sarter et al., 1997). *Automation surprises* occur when the system acts in unanticipated ways or fails to act at all. When this happens, the operator is left to ask, "What is the automation doing? Why is it doing that? What will it do next?" (Wiener, as cited in Woods, 1996, p. 117).

The problem of *automation surprises* is one of cooperation and observability (Christoffersen and Woods, 2002). In this respect, both the system performance and the operator performance may be quantified through research. This research aims to increase cooperation and with increased cooperation, increased overall performance and reliability, and decreased operator frustration. Research in this area may be divided into several areas: an operator's mental workload (workload),  an operator awareness of the current situation (situation awareness), an operator loss of skill (skill decrement), and an operator's trust in the automation to perform the assigned task (trust).

7.2 WORKLOAD

Derrick (1988) describes mental workload "as the difference between the information processing capacity available to the operator and the capacity required for criterion task performance at any given time" (p. 96). For example, a task that requires an operator to remember nine or more large numbers during a conversation with another operator who is about to perform a handoff of the system to the first operator, describes a workload problem. It is simply beyond the working memory capacity of most humans to accomplish this sequence of tasks satisfactorily. Workload refers to the number of and the perceived difficulty of the operator's tasks while operating the system. However, workload is the operator's subjective and individual perception. As Parasuraman, Sheridan, and Wickens (2008) note, "Two people performing the same task can generate identical behavioral and performance output, yet one person may have plenty of attention left to allocate to concurrent tasks, whereas the other does not" (p. 146).

Parasuraman et al. (2008) recount that in 1979, the Federal Aviation Administration (FAA) asked Sheridan and Simpson to investigate the possibility of reducing a manned aircraft flight crew from three officers to two officers. After months of observing the tasks assigned to various members of the three-officer flight crew, Sheridan and Simpson (1979) developed a workload rating scale. The workload rating scale asked officers how difficult they perceived their tasks to be and how much mental concentration or full attention was required for each task. The resulting rating scale was analogous to the Cooper–Harper Handling Qualities Scale. Further investigation using this scale resulted in their recommendation that a reduction in crew members would not significantly overburden the remaining crewmembers' workload.

Workload can be measured with the goal of predicting future operator workload requirements (Sheridan and Simpson, 1979), or as current system imposed workload, or as current operator experienced workload (Wickens and Hollands, 2000). In each case, workload may be measured by subjective or objective (i.e., psychophysiological) assessment. Objective assessments provide a continuous measure of workload but require special equipment (discussed later). Subjective assessments are easier to administer either during or after the task(s) but require the participating operator to interrupt the task or recall the task.

7.2.1 SUBJECTIVE WORKLOAD ASSESSMENTS

Some of the more commonly used subjective workload assessments are the National Aeronautics and Space Administration Task Load Index (NASA TLX) (Hart and Staveland, 1988), the Subjective Workload Assessment Technique (SWAT) (Reid et al., 1981), and the Modified Cooper–Harper Handling Qualities Scale (MCH) (Casali and Wierwille, 1983). Verwey and Veltman (1996) and Hill et al. (1992) provide a comparison of several assessments. Each of the assessments is a subjective measure of perceived workload developed for a specific, domain-related goal. Because of the subjective nature of the measure and variability in human attention, subjective workload measures are most useful when comparing UASs overall or a particular task within one UAS. Hill et al. (1992) note that the NASA TLX may provide the greatest number of dimensions with the best resolution per dimension.

A subjective workload measure will have several sections (sometimes referred to as dimensions) that the operator can complete either during or immediately after using a system. The number of dimensions and specificity of the measurement depends on which assessment tool is chosen. The NASA TLX (Hart and Staveland, 1988) measures the dimensions of physical effort, mental effort, temporal dimension, performance, effort, and frustration along a 7-point scale. This assessment has been modified for use in comparing workload across several UAS platforms by New Mexico State University (Elliott, 2009). The NMSU UAS TLX asks the operator for the names of the UASs that are being evaluated and which phases of flight the operator wishes to assess for each UAS. The program saves the results to a text file, which then may be imported into a database program. An operator may assess several different UASs in preparation for choosing a platform for purchase or for mission planning.

7.2.2 Objective Workload Assessments

Also called psychophysiological measures (Backs et al., 1994), objective workload measures typically require the operator to be fitted with special equipment (e.g., electrodes or an eye tracker). The measures include but are not limited to electroencephalogram (EEG) (Gundel and Wilson, 1992; Kramer, 1991; Sterman and Mann, 1995; Wilson and Eggemeier, 1991), event-related potential (ERP) (Humphrey and Kramer, 1994), heart rate variability (Wilson and Eggemeier, 1991), pupil dilation, and eye blink/fixation/gaze duration (Gevins et al., 1998; Gevins and Smith, 1999; Nikolaev et al., 1998; Russell and Wilson, 1998; Russell et al., 1996; Wilson and Fisher, 1995; Wilson and Russell, 2003a, 2003b). The psychophysiological measures provide continuous data during a task and also may be useful for adaptive automation events as discussed further in this chapter (Wilson, 2001, 2002; Wilson and Russell, 2003b, 2007).

Often, operators will choose not to use an automated aid because of the change required to operator workload. Although much of the automated aid research has been in manned aircraft, similar results have been found in other partially automated systems. Parasuraman and Riley (1997) report that the choice to not use an automated aid is related to pilot workload. Pilots choose not to use an aid when the aid appears at precisely the time when their workload is the greatest. Pilots do not have the time to set up the automation in addition to flying the plane; the cognitive effort to adjust the aid to the situation in progress often supersedes the workload reduction benefit. Kirlik (1993), as cited in Parasuraman and Riley (1997), found that when factors such as cognitive overhead were put into a Markoff model analysis to identify the optimal strategy in automation use, conditions favored the use of manual over automated control.

7.3 SITUATION AWARENESS

As Endsley (1996) states, "Situation awareness (SA) is a person's mental model of the world around them," or "the perception of the elements in the environment within a volume of time and space, the comprehension of their meaning and the projection of their status in the near future" (p. 164). Automation effects SA by changing the operator's role from actively controlling the system to passively monitoring the system (Endsley, 1996). This change impacts an operator's understanding of the system because of the inherent complexity associated with automation along with other factors that contribute to out-of-the-loop performance decrements (Endsley, 1996). The absence of manual system control contributes substantially to a loss of SA. As Parasuraman et al. (2000) state, "Humans tend to be less aware of changes in environmental or system states when those changes are under the control of another agent (whether that agent is automation or another human) than when they make the changes themselves" (Parasuraman et al., 2000, p. 290).

As stated before, workload and SA share an inverse relationship during an operator's use of automation. As automation takes over and reduces operator workload, operators lose SA. The loss can be described by the various levels of SA (Endsley, 1996, p. 2):

Level 1 SA—Perceiving critical factors in the environment
Level 2 SA—Understanding what those factors mean when integrated with a person's goals
Level 3 SA—An understanding of what will happen with the system in the near future

Additional sensors and better interface design may help alleviate the issue. But, if the operator has an incorrect or incomplete mental model of how the automation completes the system task, or the operator has little or no involvement in the task, or if the system should need the operator's intervention, then the out-of-the-loop operator unfamiliarity will affect operator's skill during intervention.

7.4 SKILL DECREMENT

Decreased workload and then decreased SA leads to skill decrement. As Parasuraman et al. (2000) state, "There is a large body of research in cognitive psychology documenting that forgetting and skill decay occur with disuse" (p. 291). If an operator is no longer a viable part of a task that has become automated, the task will not be practiced. This loss of practice along with an incomplete mental model of what the system is doing leaves an operator unable to successfully intervene during automation failures. Hence, giving the operator a monitoring role as their primary task creates a skill decrement problem.

Although these comments refer to research in manned flight, the findings are also true of UAS operators who rely on the system for sensory information. McCarley and Wickens (2005) state that "as compared to the pilot of a manned aircraft, thus, a UAS operator can be said to perform in relative 'sensory isolation' from the vehicle under his/her control" (p. 1). Operators rely on the interface design to communicate information from the visual and proprioceptive sensors. If sensory communication is lacking, Level 1 SA suffers, and operators must fill in the missing information with their best guess. In an uncertain environment, a best guess impacts Level 2 and Level 3 SA, workload and trust.

7.5 TRUST

Lee and See (2004) state that operators often interact with automation as they would interact with a human. In this regard, trust is related to emotion and operator attitudes regarding the systems' ability to fulfill promised task commitments. Operator attitudes are built over time and experience with the system or similar systems (Nass et al., 1995; Sheridan and Parasuraman, 2006). Communication, automation transparency, and automation reliability all contribute to building trust in systems as they do in society.

In society, trust is in part built by following mutually agreed upon rules of communication and etiquette such as Grice's maxims of communication (Grice, 1975). Miller et al. (2004) created "automation etiquette guidelines" for operator–system communication based on Gricean maxims. Parasuraman and Miller (2004) found that the automation etiquette guidelines did increase human trust of automation. In Sheridan and Parasuraman (2006, p. 103), they state that "the etiquette guidelines were powerful enough to overcome low automation reliability: performance in the low-reliability/good-etiquette condition was almost as good as ... that in the high reliability/poor etiquette condition."

In addition to etiquette to enhance communication, Klein et al. (2004, pp. 92–94) suggest that automation should be a "team player." They suggest 10 challenges to designers for making automation "team" friendly:

1. Maintaining a common grounding between operator and system—Notify each team member of impending failures
2. Model each others' intents and actions through shared knowledge, goals, and intentions
3. Predictability
4. Amenability to direction—Autonomous behavior that is consistent and an operator's option to redesignate tasks according to a systems' behavior
5. Status and intentions made obvious
6. Nuance detection/observability—An ideal system would understand pauses, rapid typing, and nonverbal human signals
7. Goal negotiation—Communicate situation change and goal revision
8. Planning and autonomy collaboration
9. Attention saliency signals—Identify the most important information being communicated
10. Cost control—Maintain action conservation

Klein et al. (2004) suggests solid but sometimes untenable goals for better collaboration between operators and systems. As Sheridan and Parasuraman suggest (2006), a system designer can improve the communications with an operator, improve trust, reliability, and operator acceptance.

7.5.1 Reliability

Lee and See (2004) also state that "trust depends on an evaluation made under uncertainty, in which decision makers use their knowledge of the motivations and interests of the other party to maximize gains and minimize losses" (p. 62). In this respect, trust is a reciprocal decision based on decision making under uncertainty, as in expected utility theory (e.g., as described by Kahnemann and Tversky, 1979). Operators, who are uncertain about automated task performance, trust automation designers and the automated system to behave in the most expeditious and beneficial manner, thereby maximizing benefits to the operator while minimizing loss. When this promise is not achieved, operator trust in the automated system suffers. Wickens and Dixon (2007) suggest that a system that has a reliability of 70% or less is "worse than no automation at all" (p. 201). They go on to state that operators will protect the performance of tasks from automation that is seriously imperfect.

Parasuraman and Riley (1997) make a compelling case that "system designers should be concerned about use, misuse, disuse, and abuse of automation based on distrust and over-trust as well as on workload and other factors" (p. 249). Use, misuse, disuse, and abuse of automation refer to the tendency of a human operator to refuse to use (disengage) the automation, to overrely (fail to monitor) on the automation, disuse or ignore the automated alarms, and design abuse or the tendency for designers to automate tasks without consideration of the effect on operator performance. The article goes on to outline possible factors implicated in each of these outcomes and has inspired HF researchers to delve into the factors that contribute to each outcome.

"Inappropriate reliance associated with misuse and disuse depends, in part, on how well trust matches the true capabilities of the automation" (Wicks, Berman, and Jones, 1999, p. 99). Rice (2009) has proposed that the two types of automation errors (false alarms and misses) elicit two types of operator responses (compliance and reliance). In the study, participants were given several levels of system reliability. One system was more likely to report a target when there was no target (false alarm prone system). The second system was more likely to report no target when there was a target present (miss prone system). Participants made judgments while operating each of the two systems. The false alarm prone system elicited different judgments from participants than the missed prone system. This suggested that human operators are biased in their judgmxents based on the type of imperfection in the automation. Rice (2009) states that "there is much data that suggest that false alarms are more damaging overall than misses (see, for example, Bliss, 2003) and that the two types of errors differentially affect operator trust (see, for example, Dixon and Wickens, 2006; Maltz and Shinar, 2003; Meyer, 2001, 2004; Wickens and Dixon, 2007)" (p. 305).

7.6 TYPES AND LEVELS OF AUTOMATION

7.6.1 Types of Automation

Automation is often thought of as an all-or-none proposition. But, as the variety in current UASs demonstrates, automation occurs at many levels and in many varieties. McCarley and Wickens (2005) point out that controls in UASs range from a UAS controlled by a manual system of stick and rudder to a UAS controlled by a ground control station through which an operator can preplan a mission and make changes in real time, to a fully automated control system that flies to preplanned coordinates and performs preprogrammed tasks. Within this range, automation can be divided into two categories: designer-created binary automation (static automation) or context-dependent automation (adaptive automation). Static automation is hard-wired into the system. The system designer chooses whom (the system or the operator) and how (manual or automatic) a task will be performed. The designer may allow the operator to override the automation or configure it to meet a changing situation. Adaptive automation is called up by an operator event. The event may be explicit (a request for aid) or the event may be implicit (tied to operator workload) or a situational event (takeoff speed). Adaptive automation is characterized by the ability to turn itself on in connection with a system or an operator event. Each type of automation confers benefits but at a cost. Parasuraman et al. (1992) proposed a series of questions a designer might consider when designing either type of automation. As Morrison (1993) states, "Adaptive automation has the potential to solve many problems that are created by or not addressed by conventional automation."

7.6.1.1 Adaptive Automation (AA)

"Experimental and theoretical research in adaptive aiding that began in 1974, [was] motivated to a great extent by concern for how humans and artificially intelligent systems should interact" (Rouse, 1988, p. 432). Rouse (1988) continues by noting that research and design efforts at the time began with "hobby shopping" and not with considerations of alternative functions designed to relieve an overburdened operator. Early research during this time includes manned aircraft adaptive automation (AA). Chu and Rouse (1979) found that response time in flight tasks showed a 40% reduction with AA.

Since Rouse and others' early work (see Inagaki, 2003; Scerbo, 1996, 2007, for a review), subsequent researchers have found similar effects (Parasuraman et al., 2000). These include studies by Parasuraman et al. (1992, 1993, 1996, 1999), Scallen et al. (1995), Hancock and Scallen (1996), Hilburn et al. (1997), Kaber and Riley (1999), and Moray et al. (2000). Issues such as unbalanced workload, loss of SA, and skill loss can be addressed successfully by implementing AA (Parasuraman et al., 2000).

Current AA implementations address issues of high workload, loss of SA, and skill decline. Kaber and Endsley (2004) and Parasuraman et al. (1996) describe AA that invokes when operator workload is unnecessarily high and when it is unnecessarily low. In both cases, operator stress is reduced and overall system performance increases. Parasuraman and Wickens (2008b) describe a study in which Wilson and

Russell (2003a) used psychophysiological data and an artificial neural network to discern states of low and high workload. When high workload was detected, the AA was invoked to perform low-level tasks. An overall improvement in system performance was found. In 1989, the Naval Air Development Center (NADC) proposed that aircraft automation, which is static, has led to operator difficulties. They suggested that a dynamic approach to automation be explored.

AA may be invoked through several different types of events. Parasuraman et al. (1992) outlined several categories for the creation of an AA-linked event: critical event logic, dynamic assessment of operator workload, dynamic operator psychophysical assessment, and performance models. Critical event logic is the simplest to implement. It ties the invocation of the AA to specific tactical events prescribed by doctrine or procedure manuals. Barnes and Grossman (1985) offer a review of the levels of events and specifics of this approach. Critical event logic is based upon the assumption that operator workload consistently increases after a critical event. Dynamic assessment of operator workload is a continuous online monitoring of operator characteristics during work. A performance measure may be used to create events that invoke AA with the purpose of maintaining a median level of operator workload. Dynamic operator psychophysical assessment is the same as dynamic operator assessment, but psychophysical methods of workload (e.g., ERP, pupil dilation) are used as a continuous measure of operator workload. AA is invoked when predetermined parameters have been exceeded. Finally, performance models sometimes are used to model predictor levels of operator workload and system resources. Once a performance model threshold is exceeded, the AA is invoked (e.g., the system requires the operator to perform several tasks simultaneously, with each task requiring a different sensory modality). Performance models include but are not limited to optimal mathematical models such as signal detection theory, inferencing models, central executive models, and abductive approaches (see, for example, Wickens multiple resource theory; Wickens 1979, 1984).

However, AA is not without unique challenges as an event-invoked automation. Parasuraman et al. (2000) suggest that if a critical event does not occur, AA is not invoked. Billings and Woods (1994) cite the perceived unpredictability of the system by operators. One solution may be to allow only the operator to invoke the AA (Parasuraman and Wickens, 2008a). Other solutions might be to create a common communication platform between the AA and the operator so that tasks might be delegated as a supervisor or coach delegates to players (Parasuraman et al., 2005).

7.6.1.2 Adaptive Automation Implementations

AA theory and research has recently been implemented in automation development efforts. Parasuraman and Wickens (2008a) mention several research studies but only a single implementation: Rotorcraft Pilot's Associate (Dornheim, 1999). Pilot's Associate assists helicopter operators and has successfully passed in-flight performance tests. Several other projects, at the time of publication, seem to be in the test and evaluation phases of development.

Playbook, by Miller, Goldman, and Funk (2004), is an adaptive automation system built on the metaphor of a sports playbook. Playbook, an AA, creates a communication platform for use with the subsystems of plan delegation, constraint avoidance, or stipulations. It integrates with planning expert systems and with variable autonomy control systems, and proposes to cover hypothetical situations and contingencies. With the playbook metaphor, operators summon and compile plans of action, which include goals, constraints, stipulations, and policies. The AA then checks the viability of the operator request and issues commands. Plays may include tasks such as sustained surveillance of an area, tracking a target, and watch perimeter. The system operator has prior knowledge of the plays available, constraints, expectations, and so on.

Another AA implementation is incorporated into RoboFlag (Squire et al., 2006). RoboFlag is a computerized version of the children's game capture the flag. Currently, RoboFlag is used in experimental laboratories to test AA effects on operator performance in various types of task paradigms (Squire et al., 2006).

7.6.1.3 Autonomy

Whereas AA is dynamic and flexible, traditional automation is static; total automation or autonomy is neither. Autonomy or full automation of a UAS proposes a strong artificial intelligence (AI) approach to automation. For decades, cognitive researchers have struggled with the strong AI stance, which proposes that human cognition may be replicated by a machine. Designers posit success with a strong AI approach to automation and propose a future of fully automated, intelligent UASs.

7.6.1.3.1 Autonomous Implementations

Most notably, Pettersson and Doherty (2004) with the Wallenberg Laboratory for Information Technology and Autonomous Systems (WITAS) at Linköping University in Sweden have done design work in autonomous UAS. Specifically, DyKnow is a general knowledge-processing framework that serves on top of existing middleware platforms connecting "knowledge representation and reasoning services, grounding knowledge in sensor data and providing uniform interfaces for processing and management of generated knowledge and object structures" (Heintz and Doherty, 2004).

7.6.2 Levels of Automation (LoA)/Human-Centric Taxonomies

Several researchers propose the use of taxonomies or levels of automation for research, evaluation, testing, and design. Among the most prominent human-centric taxonomies are those suggested by Sheridan and Verplank (1979); Parasuraman et al. (2000); Ntuen and Park (1988); Endsley (1987); and Endsley and Kaber (1999). The human-centered taxonomies are useful for isolating operator and system performance issues and defining what the automation can and should be doing in terms of human cognitive performance (Wickens, 2008a). Parasuraman et al. (2000) have one of the most widely used taxonomies. They propose that automation can be separated into four classes of functions: "(1) information acquisition; (2) information analysis; (3) decision and action selection; (4) action implementation" (Parasuraman et al., 2000, p. 288). These classes mirror human cognition when the same system task is performed manually. Endsley and Kaber (1999) have created a similar, 10-level taxonomy that is intended to describe cognitive and psychomotor tasks in domains that require real-time control.

Endsley and Kaber's LoA is particularly useful because they have demonstrated the impact automation has on the operator's ability to assume manual control at the different taxonomic levels (1999). In their study, they asked participants to monitor several targets on a screen at the different LoAs. At intervals throughout, participants were to report on their SA and workload. As participants served as operators, several automation failures occurred and their ability to recover the system manually was assessed. When operator performance in each of the different LoAs were then compared, Endsley and Kaber found that automation did affect performance at different LoAs with higher LoAs resulting in decrements in operator performance and lower level LoAs helped operator performance.

7.7 TECHNOCENTRIC TAXONOMIES

In the unmanned systems community, a growing need to define autonomy as it pertains to unmanned systems has been challenging. Government agencies and contractors have struggled to develop a general LoA from which useful performance measures can be deduced. The Air Force Research Laboratory (AFRL), NASA, and others have developed program-specific LoA. An overview of their respective taxonomies or levels of autonomy are described next.

7.7.1 Air Force Research Laboratory (AFRL)

AFRL was directed to develop a national intelligent autonomous UAV control metric according to (Clough, 2000). This effort focuses on the fixed-wing vehicle initiative (FWV). Some key distinctions were made in this effort regarding automatic and autonomous as well as autonomy and intelligence. It is stated that an autopilot is automatic in that it stays on the course chosen, whereas an autonomous guidance system decides which course to take and then stays on it. Autonomy is defined in this effort as "the ability to generate one's own purposes without any instruction from outside" (Clough, 2002, p. 1) and "having free will" (Clough, 2002, p. 1). Clough also stated, "Intelligence … is the capability of discovering and using it to do something" (2002, p. 1). The main objective was to know how well the UAS did its assigned task and not on the performance of the abilities that allowed the system to complete its task.

AFRL's search for current autonomy metric led them to Los Alamos National Laboratory's Mobility, Acquisition, and Protection (MAP) and Draper Laboratory's Three Dimensional Intelligence Space. AFRL's search did not provide metrics that could be used directly so it decided to integrate the most useful information from its findings. The autonomous control level chart was created (Clough, 2002) from their efforts.

7.7.2 NASA

NASA has concluded that for its vision of space exploration to be fulfilled, its systems will need to become more autonomous and have higher levels of automation. NASA's approach is to define the levels of autonomy and automation needed for a mission and design the system to those requirements. This is different from the other approaches presented here. The two questions NASA is seeking to answer are: What is the right balance of ground versus onboard authority (autonomy)? What is the right balance of human versus computer authority (automation)? NASA has developed the Function-Specific Level of Autonomy and Automation Tool (FLOAAT) to facilitate its system requirements development. This tool uses two scales: one for autonomy and the other for automation (Proud and Hart, 2005).

The automation scale has five levels of automation for each of the four stages of decision making, as specified by the OODA Loop. The OODA loop, otherwise known as the Boyd cycle, has an observe, orient, decide, and act stage developed by USAF Colonel John Boyd (Brehmer, 2005). The lowest level of automation states that all data monitoring, calculation, decisions, and tasks are executed by the ground station. The highest level of automation moves all data monitoring, calculation, decisions, and task execution to the onboard system. The levels between these have a linear transition from completely ground-based control to onboard automation (Proud, 2005).

The autonomy scale has eight levels of autonomy for each stage of the OODA loop. The lowest level of autonomy for the observe stage is done completely by the human as are the other orient, decide, and act stages. At the highest level of autonomy, the observe stage does all the data mining without any assistance from the human. Similarly the orient, decide, and act stages all are done without any human intervention.

These scales, along with a questionnaire, allow domain-specific experts to evaluate the level of autonomy needed in developing requirements for a system or provide criteria for specifying the current autonomy level of a system. In the latter case, the NASA levels of autonomy would be comparable to the other levels of autonomy (Proud, 2005).

7.7.3 Others

Other attempts have been made to define autonomy and their associated levels for specific programs. As part of the unmanned combat air vehicle (UCAV) program, the Defense Advanced Research Projects Agency (DARPA), U.S. Air Force, and Boeing developed a top-level view of autonomy (see Figure 7.1) Also, as part of the

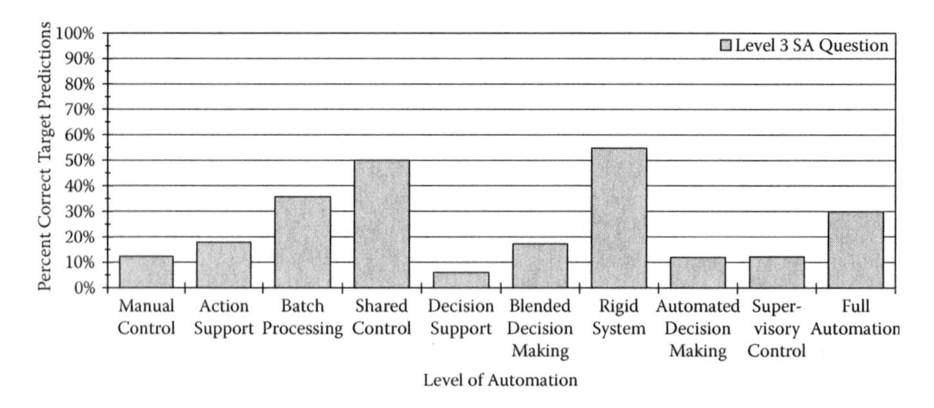

FIGURE 7.1 A top-view level of autonomy.

U.S. Army Future Combat Systems Program, a more detailed LoA scale was created as compared to the UCAV LoA (Committee on Autonomous Vehicles in Support of Naval Operations, National Research Council, 2005).

The U.S Army Future Combat System (FCS) LoA are broken into 10 levels ranging from remote control (level 1) to fully autonomous (level 10) (see Figure 7.2). At each level in the autonomy taxonomy, a description of the level, observation perception/SA, decision making, and capability are provided along with an example scenario for each level (Committee on Autonomous Vehicles in Support of Naval Operations, National Research Council, 2005).

7.8 NATIONAL INSTITUTE OF STANDARDS AND TECHNOLOGY

The National Institute of Standards and Technology (NIST) has been working on the Autonomy Levels for Unmanned Systems (ALFUS) framework since 2003 (Figure 7.3). The ALFUS framework was developed through collaboration of military and civilian practitioners from the fields of unmanned ground vehicles (UGV), unmanned air vehicles (UAV), unmanned surface vehicles (USV), and unmanned undersea vehicles (UUV). The major objectives of ALFUS were to provide standard terms and definitions of unmanned systems' autonomous capabilities and provide metrics, processes, and tools for evaluating the unmanned systems' autonomy.

Within the ALFUS framework, autonomy is defined as "a UMS's own ability of integrated sensing, perceiving, analyzing, communicating, planning, decision-making, and acting/executing, to achieve its goals as assigned." The LoA given to an unmanned system is evaluated by three aspects. First, the mission complexity (MC) is the measure of difficulty of the mission to be performed. Second, the environmental complexity (EC) is the measure of difficulty of the environment that the mission is conducted. Third, the human interaction (HI) is the measure of human interaction during the mission. During the development of ALFUS, the FCS also was looking into defining metrics for autonomy. Therefore, metrics of MC and EC were focused on military applications. A major difficulty in defining LoA is that each aspect rarely is treated independently. Therefore, developing standard metrics for MC, EC, and HI has proven to be difficult and is done on a test-by-test basis. The reader is referred to ALFUS for further information (Huang et al., 2007).

In defining the LoA, metrics must be defined that encompass the complexities of the test. These metrics will be scored, weighted, and summed to a composite score and a LoA can be assigned to the system. As mentioned earlier, defining these metrics is a difficult task.

7.8.1 Level of Autonomy 0

The lowest level of autonomy is completely controlled by the human. The human has direct control over primitive locomotion functions, and the system has no ability to alter. Level 0 is analogous to driving a remote-controlled (RC) car or plane. The human directly controls the actuator speed and position.

Level	Observe	Orient	Decide	Act
5	The data is monitored onboard without assistance from ground support.	The calculations are performed onboard without assistance from ground support.	The decision is made onboard without assistance from ground support.	The task is executed onboard without assistance from ground support.
4	The majority of the monitoring will be performed onboard with available assistance from ground support.	The majority of the calculations will be performed onboard with available assistance from ground support.	The decision will be performed onboard with available assistance from ground support.	The task is executed onboard with available assistance from ground support.
3	The data is monitored both onboard and on the ground.	The calculations are performed both onboard and on the ground.	The decision is made both onboard and on the ground and the final decision is negotiated between them.	The task is executed with both onboard and ground support.
2	The majority of the monitoring will be performed by ground support with available assistance onboard.	The majority of the calculations will be performed by ground support with available assistance onboard.	The decision will be made by ground support with available assistance onboard.	The task is executed by ground support with available assistance onboard.
1	The data is monitored on the ground without assistance from onboard.	The calculations are performed on the ground without assistance from onboard.	The decision is made on the ground without assistance from onboard.	The task is executed by ground support without assistance from onboard.

FIGURE 7.2 (a) NASA level of automation. (b) NASA level of autonomy. (continued)

Level	Observe	Orient	Decide	Act
8	The computer gathers, filters, and prioritizes data without displaying any information to the human.	The computer predicts, interprets, and integrates data into a result which is not displayed to the human.	The computer performs final ranking, but does not display results to the human.	Computer executes automatically and does not allow any human interaction.
7	The computer gathers, filters, and prioritizes data without displaying any information to the human. Though, a "program functioning" flag is displayed.	The computer analyzes, predicts, interprets, and integrates data into a result which is only displayed to the human if result fits programmed context (context-dependant summaries).	The computer performs final ranking and displays a reduced set of ranked options without displaying "why" decisions were made to the human.	Computer executes automatically and only informs the human if required by context. It allows for override ability after execution. Human is shadow for contingencies.
6	The computer gathers, filters, and prioritizes information displayed to the human.	The computer overlays predictions with analysis and interprets the data. The human is shown all results.	The computer performs ranking tasks and displays a reduced set of ranked options while displaying "why" decisions were made to the human.	Computer executes automatically, informs the human, and allows for override ability after execution. Human is shadow for contingencies.
5	The computer is responsible for gathering the information for the human, but it only displays non-prioritized, filtered information.	The computer overlays predictions with analysis and interprets the data. The human shadows the interpretation for contingencies.	The computer performs ranking tasks. All results, including "why" decisions were made, are displayed to the human.	Computer allows the human a context-dependant restricted time to veto before execution. Human shadows for contingencies.
4	The computer is responsible for gathering the information for the human and for displaying all information, but it highlights the nonprioritized, relevant information for the user.	The computer analyzes the data and makes predictions, though the human is responsible for interpretation of the data.	Both human and computer perform ranking tasks, the results from the computer are considered prime.	Computer allows the human a pre-programmed restricted time to veto before execution. Human shadows for contingencies.
3	The computer is responsible for gathering and displaying unfiltered, unprioritized information for the human. The human is still the prime monitor for all information.	Computer is the prime source of analysis and predictions, with human shadow for contingencies. The human is responsible for interpretation of the data.	Both human and computer perform ranking tasks, the results from the human are considered prime.	Computer executes decision after human approval. Human shadows for contingencies.
2	Human is the prime source for gathering and monitoring all data, with computer shadow for emergencies.	Human is the prime source of analysis and predictions, with computer shadow for contingencies. The human is responsible for interpretation of the data.	The human performs all ranking tasks, but the computer can be used as a tool for assistance.	Human is the prime source of execution, with computer shadow for contingencies.
1	Human is the only source for gathering and monitoring (defined as filtering, prioritizing, and understanding) all data.	Human is responsible for analyzing all data, making predictions, and interpretation of the data.	The computer does not assist in or perform ranking tasks. Human must do it all.	Human alone can execute decision.

FIGURE 7.2 (continued) (b) NASA level of autonomy.

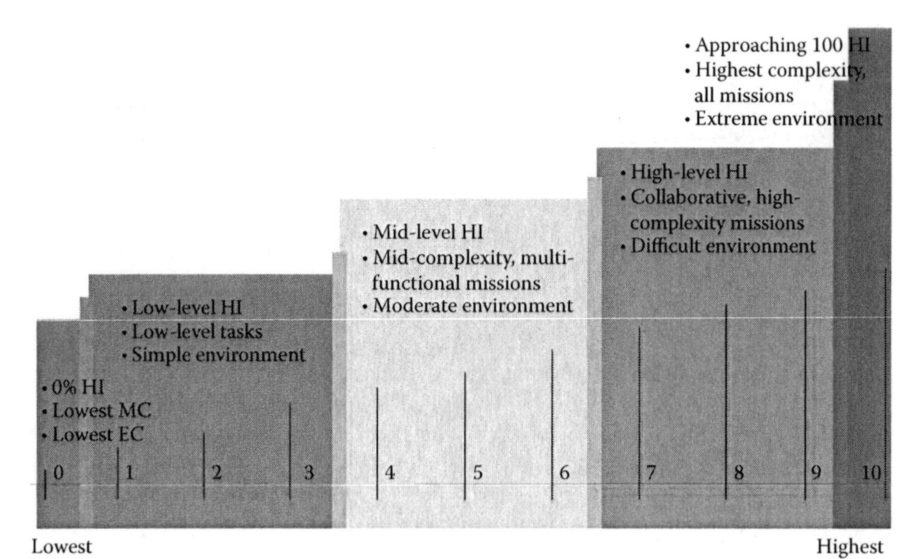

FIGURE 7.3 UCAV LoA.

7.8.2 Level of Autonomy 1–3 (Low LoA)

Systems that have very little internal SA and have low levels of automation for subsystem tasks would have a 1 to 3 LoA for the MC aspect. The EC aspect is in this range if the environment exhibits very benign features that allow the mission to have a high success rate. These environments have static and simple features that the UAS can utilize. The human interaction is the main component in these systems. Humans will interact with UASs the majority of the time by commanding the action that the UAS will execute. Extending the RC car/plane example, the operator will no longer have direct actuator control of the UAS. The subsystem automation of the velocity control will allow the operator to specify a direction or waypoint, and allows the UAS to execute the command. The UAS will control its velocity and pose as dictated by the operator using its minimal internal SA.

7.8.3 Level of Autonomy 4–6 (Mid LoA)

The systems that exhibit a mid LoA will interact with the human about 50% of the time. The human provides the UAS with goals, and the system decides how to execute those goals. Before the UAS can execute, the human must give final approval. There is a significant difference between the UAS at a low LoA and a mid LoA in the high aspect. There is a linear transition from the low LoA for the MC and EC to the mid LoA. The biggest difference in the MC aspect is the limited real-time planning ability of the UAS, which is not present at the low LoA. The EC has transitioned from a low-risk and static environment to a mid-risk and understandable dynamic environment. A simple example would be to have a UAS given a command to find object A. The system would propose a plan of execution to accomplish the goal and the human would approve it. The environment may be difficult to transverse and may have dynamic objects or the object itself may be dynamic.

7.8.4 Level of Autonomy 7–9 (High LoA)

UASs at a high LoA have very little interaction with the human. The system no longer needs approval from the human for execution of the goals. The UAS informs the human but will carry out the plans unless there is human intervention. The EC has a high risk of failure with great uncertainty and an understandable highly dynamic environment. The MC is more focused on teams of manned and unmanned systems. The systems will have high fidelity SA with real-time planning capable of high adaption and complex decision making. These are systems that are capable of performing very complicated tasks to accomplish complicated goals.

7.8.5 Beyond Level 10

At this LoA, the UAS is considered to perform at a human level. The human interaction is no longer for oversight but for acquiring information from appropriate individuals to accomplish the mission. The MC and EC are of the highest difficulty. The UAS is considered omniscient and able to overcome situations with the highest uncertainty and the lowest probability of success (Huang et al., 2007).

At present, there is no agreed upon definition of intelligence for humans or machines. We can generalize ALFUS's definition of autonomy to be the ability of the system to achieve its goals as assigned. This would suggest that UASs exhibit a minimum level of competence much like adults exhibit an expected minimum level of competence for subjects such as reading or math. However, humans have the unique ability to perform abstract judgment and reasoning tasks in undefined or ill-defined circumstances. So, it is unclear whether systems at a LoA 10 can be considered to be of human-level intelligence.

The major difficulty in characterizing a UAS's LoA is defining meaningful performance metrics for unmanned systems. This is why a working group has been initiated to examine such metrics. The Performance Measures Framework for Unmanned Systems (PerMFUS) goal is to provide the community a way of capturing the UAS performance from a technical and operational perspective (Huang et al., 2009). The Society of Automotive Engineers Aerospace Systems Division (SAE AS-4) has an Unmanned Systems Technical Committee that also has been examining these performance metrics. The mission of SAE AS-4 is to address all facets of unmanned systems with a primary goal of publishing open system standards and architectures that the military, civil, and commercial communities can leverage. AS-4D subcommittee's primary objective is to provide terms, definitions, and measures for accessing the performance of unmanned systems.

7.9 CONCLUSION

In this chapter, human-centric and technocentric taxonomies have been reviewed. Both are very useful in defining design and operating functional specifications. Although a universal taxonomy might seem advantageous, taxonomies that address the technological aspects of design as well as the human aspects of design in each domain offer the most advantages.

In addition to taxonomies, the chapter reviewed other issues in UAS automation: the tradeoff between operator workload and operator situation awareness, the importance of system–operator communication, and the importance of an accurate operator mental model of the system. Finally, the chapter reviewed the importance of system reliability in operator trust and how reliability can change operator bias. Although many autonomous UAS projects (systems that purport to have no human in the loop) are being developed, the same issues of situation awareness, workload tradeoffs, and system reliability apply. Except these issues apply within the system itself. Unfortunately, the complexities introduced within the system will make failures and imperfections more difficult to pinpoint. Recent literature on autonomous systems supports this claim and cites the inability of autonomous systems to create consistent object representations from a variety of sensory information. For example, as sensors represent an object such as a car to the system, one set of sensors may represent the object as permeable. Yet, another set of sensors represents the object as impermeable. Designers will not only have to continue to address the issues of system communication with an operator (e.g., in case of system failure) but they will have to address the same issues of trust, situation awareness, workload, and skills within the autonomous system itself.

In conclusion, "there is a belief among many automation engineers that one can eliminate human error by eliminating the human operator. To the extent a system is made less vulnerable to operator error; it is made more vulnerable to designer error and given that the designer is also human, this simply displaces the locus of human error. In the end, automation is really human after all" (Sheridan and Parasuraman, 2006, p. 124).

DISCUSSION QUESTIONS

7.1 What are the advantages and disadvantages of adaptive automation compared to static automation?

7.2 Human factors research suggests that automation affects UAS operators in what ways?

7.3 Choose a taxonomy and then describe the different levels. Can you think of an example of automation for each level?

7.4 What are the advantages and disadvantages to having a general taxonomy?

7.5 How would one quantify a system as operating at a specific level within a taxonomy?

REFERENCES

AS-4 Unmanned Systems Steering Committee. (n.d.). Retrieved June 9, 2010, from Society of Automotive Engineers (SAE), http://www.sae.org/servlets/works/committeeHome.do?comtID=TEAAS4.

Backs, R. W., Ryan, A. M., and Wilson, G. F. 1994. Psychophysiological measures of workload during continuous manual performance. *Human Factors* 36: 514–531.

Barnes, M., and J. Grossman. 1985.*The Intelligent Assistant Concept for Electronic Warfare Systems.* China: NWC.

Billings, C. E., and D. D. Woods. 1994.Concerns about adaptive automation in aviation systems. In *Human Performance in Automated Systems: Current Research and Trends*, ed. M. Mouloua and R. Parasuraman, 24–29. Hillsdale, NJ: Erlbaum.

Bliss, J. 2003. An investigation of alarm related accidents and incidents in aviation. *International Journal of Aviation Psychology* 13: 249–268.

Brehmer, B. 2005. The dynamic OODA loop: Amalgamating Boyd's OODA loop and the cybernetic approach to command and control. 10th International Command and Control Research and Technology Symposium— The Future of C2.

Casali, J. G., and W. W. Wierwille. 1983. A comparison of rating-scale, secondary-task, physiological, and primary-task workload estimation techniques in a simulated flight task emphasizing communications load. *Human Factors* 25:623–641.

Christoffersen, K., and D. D. Woods. 2002. How to make automated system team players. In *Advances in Human Performance and Cognitive Engineering Research*, vol. 2, ed. E. Salas, 1–12. Amsterdam: Elsevier.

Chu, Y., and W. B. Rouse. 1979. Adaptive allocation of decision making responsibility between human and computer in multi-task situations. *IEEE Transactions System, Man, and Cybernetics, SMC-9*: 769.

Clough, B. T. 2002. Metrics, schmetrics! How the heck do you determine a UAV's autonomy anyway? Performance Mectric for Intelligent Systems Workshop. Gaithersburg, MD.

Committee on Autonomous Vehicles in Support of Naval Operations, National Research Council. 2005. *Autonomous Vehicles in Support of Naval Operations*. Washington, DC: The National Academies Press.

Derrick, W. L. 1988. Dimensions of operator workload. *Human Factors* 30: 95–110.

Dixon, S. R., and C. D. Wickens. 2006. Automation reliability in unmanned aerial vehicle control: A reliance-compliance model of automation dependence in high workload. *Human Factors* 48: 474–486.

Dixon, S., C. D. Wickens, and J. M. McCarley. 2007. On the independence of reliance and compliance: Are false alarms worse than misses? *Human Factors* 49: 564–572.

Dornheim, M. 1999. Apache tests power of new cockpit tool. *Aviation Week and Space Technology* 46–49.

Elliott, L. J. 2009. NMSU_UAS_TLX [computer software]. Las Cruces, New Mexico.

Endsley, M. 1987. The application of human factors to the development of expert systems for advanced cockpits. In *Proceedings of the Human Factors Society 31st Annual Meeting*, 1388–1392. Santa Monica, CA: Human Factors Society.

Endsley, M. 1996. Automation and situation awareness. In *Automation and Human Performance: Theory and Applications*, ed. R. Parasuraman and M. Mouloua, 163–181. Mahwah, NJ: Erlbaum.

Endsley, M. R., and D. B. Kaber. 1999. Level of automation effects on performance, situation awareness and workload in a dynamic control task. *Ergonomics* 42: 462–492.

Gevins, A., and M. E. Smith. 1999. Detecting transient cognitive impairment with EEG pattern recognition methods. *Aviation, Space and Environmental Medicine* 70: 1018–1024.

Gevins, A., M. E. Smith, H. Leong, L. McEvoy, S. Whitfield, R. Du, and G. Rush. 1998. Monitoring working memory load during computer-based tasks with EEG pattern recognition methods. *Human Factors* 40: 79–91.

Grice, H. P. 1975. Logic and conversation. In *Syntax and Semantics*, vol. 3, Speech Acts, ed. Peter Cole and Jerry L. Morgan, 41–58. New York: Academic Press.

Gundel, A., and G. F Wilson. 1992. Topographical changes in the ongoing EEG related to the difficulty of mental task. *Brain Topography* 5: 17–25.

Hancock, P. A., and S. F. Scallen. 1996. The future of function allocation. *Ergonomics in Design* 4: 24–29.

Hart, S. G., and L. E. Staveland. 1988. Development of the NASA-TLS (Task Load Index): Results of empirical and theoretical research. In *Human Mental Workload*, ed. P. A. Hancock and N. Meshkati, 239–250. Amsterdam: North Holland.

Heintz, F., and P. Doherty. 2004. DyKnow: An approach to middleware for knowledge processing. *Journal of Intelligent and Fuzzy Systems* 15: 3–13.

Hilburn, B., P. G. Jorna, E. A. Byrne, and R. Parasuraman. 1997. The effect of adaptive air traffic control (ATC) decision aiding on controller mental workload. In *Human-Automation Interaction: Research and Practice*, ed. M. Mouloua and J. Koonce, 84–91. Mahwah, NJ: Erlbaum.

Hill, S. G., H. P. Iavecchia, J. C. Byers, A. C. Bittner Jr., A. L. Zaklad, and R. E. Christ. 1992. Comparison of four subjective workload rating scales. *Human Factors* 34: 429–440.

Huang, H.-M., E. Messina, and A. S. Jacoff. 2009. Performance Measures Framework for Unmanned Systems (PerMFUS): Initial perspective. Proceedings of the Performance Metrics for Intelligent Systems (PerMIS) 2009 Conference.

Huang, H. M., E. Messina, and J. Slbus. 2007. Autonomy Levels for Unmanned Systems (ALFUS) Framework, vol. II. NIST Special Publication 1011-II-1.0.

Humphrey, D. G., and A. F. Kramer. 1994. Toward a psychophysiological assessment of dynamic changes in mental workload. *Human Factors* 36: 3–26.

Inagaki, T. 2003. Adaptive automation: Sharing and trading of control. In *Handbook of Cognitive Task Design*, ed. E. Hollnagel, 46–89. Mahwah, NJ: Erlbaum.

Kaber, D. B., and M. R. Endsley. 2004. The effects of level of automation and adaptive automation on human performance, situation awareness and workload in a dynamic control task. *Theoretical Issues in Ergonomics Science* 5: 113–153.

Kaber, D. B., and J. M. Riley. 1999. Adaptive automation of a dynamic control task based on workload assessment through a secondary monitoring task. In *Automation Technology and Human Performance: Current Research and Trends*, ed. M. W. Scerbo and M. Mouloua, 129–133. Mahwah, NJ: Erlbaum.

Kahneman, D., and A. Tversky. 1979. Prospect theory: An analysis of decision under risk. *Econometrica: Journal of the Econometric Society* 47: 263–291.

Kirlik, A. 1993. Modeling strategic behavior in human-automation interaction: Why an "aid: can (and should) go unused. *Human Factors: The Journal of the Human Factors and Ergonomics Society* 35: 221–242.

Klein, G., D. D. Woods, J. M. Bradshaw, R. R. Hoffman, and P. J. Feltovich. 2004. Ten challenges for making automation a "team player" in joint human-agent activity. *IEEE Intelligent Systems*, 91–95.

Kramer, A. F. 1991. Physiological measures of mental workload: A review of recent progress. In *Multiple Task Performance*, ed. D. Damos, 279–238. London: Taylor & Francis.

Lee, J. D., and K. A. See. 2004. Trust in automation and technology: Designing for appropriate reliance. *Human Factors* 46: 50–80.

Maltz, M., and D. Shinar, 2003. New alternative methods in analyzing human behavior in cued target acquisition. *Human Factors, 45*: 281–295.

McCarley, J. S., and C. D. Wickens. 2005. Human factors implications of UAVs in the national airspace (Technical report AHFD-05-05/FAA-05-01). Aviation Human Factors Division, Savoy, Illinois.

Meyer, J. 2001. Effects of warning validity and proximity on responses to warnings. *Human Factors* 43: 563–572.

Meyer, J. 2004. Conceptual issues in the study of dynamic hazard warnings. *Human Factors* 46: 196–204.

Miller, C. A., Goldman, R. P. and Funk, H. B. 2004. Delegation approaches to multiple unmanned vehicle control. Proceedings of the Workshop on Human Factors of Unmanned Aerial Vehicles: Manning the Unmanned CERI, Tempe, AZ.

Moray, N., T. Inagaki, and M. Itoh. 2000. Adaptive automation, trust, and self-confidence in fault management of time-critical tasks. *Journal of Experimental Psychology: Applied* 6: 44–58.

Morrison, J. G. 1993. *The Adaptive Function Allocation for Intelligent Cockpits (AFAIC) Program: Interim Research Guidelines for the Application of Adaptive Automation.* Warminster, PA: Naval Air Warfare Center–Aircraft Division.

NADC. 1989. *Adaptive Function Allocation for Intelligent Cockpits.* Warminster, PA: Naval Air Development Center.

Nass C., Y. Moon, B. J. Fogg, B. Reeves, and C. Dryer. 1995. Can computer personalities be human personalities? *International Journal of Human Computer Studies* 43: 223–239.

Nikolaev, A. R., G. A. Ivanitskii, and A. M. Ivanitskii. 1998. Reproducible EEG alpha-patterns in psychological task solving. *Human Physiology* 24: 261–268.

Ntuen, C. A., and E. H. Park. 1988. Human factors issues in teleoperated systems. In *Ergonomics of Hybrid Automated*, by W. Karwowski, H. R. Parsaei and M. R. Wilhelm, 203–210. Amsterdam: Elsevier.

Parasuraman, R., T. Bahri, J. Deaton, J. Morrison, and M. Barnes. 1992. Theory and design of adaptive automation in aviation systems (Technical report no. NAWCADWAR-92033-60). Warminster, PA: Naval Air Warfare Center.

Parasuraman, R., S. Galster, P. Squire, H. Furukawa, and C. A. Miller. 2005. A flexible delegation interface enhances system performance in human supervision of multiple autonomous robots: Empirical studies with RoboFlag. *IEEE Transactions on Systems, Man and Cybernetics—Part A: Systems and Humans* 35: 481–493.

Parasuraman, R., and P. A. Hancock. 1999. Using signal detection theory and Bayesian analysis to design parameters for automated warning systems. In *Automation Technology and Human Performance: Current Research and Trends*, ed. M. W. Scerbo and M. Mouloua, 63–67. Mahwah, NJ: Erlbaum.

Parasuraman, R., and C. A. Miller. 2004. Trust and etiquette in high-criticality automated systems. *Communications of the ACM* 47: 51–55.

Parasuraman, R., R. Molloy, and I. L. Singh. 1993. Performance consequences of automation induced "complancency." *The International Journal of Aviation Psychology* 3: 1–23.

Parasuraman, R., M. Mouloua, and R. Molloy. 1996. Effects of adaptive task allocation on monitoring of automated systems. *Human Factors* 38: 665–679.

Parasuraman, R., and V. A. Riley. 1997. Humans and automation: Use, misuse, disuse and abuse. *Human Factors* 39: 230–253.

Parasuraman, R., T. B. Sheridan, and C. D. Wickens. 2000. A model for types and levels of human interaction with automation. *IEEE Transactions on Systems, Man and Cybernetics—Part A: Systems and Humans* 30: 286–297.

Parasuraman, R., T. B. Sheridan, and C. D. Wickens. 2008. Situation awareness, mental workload, and trust in automation: Viable, empirically supported cognitive engineering constructs. *Journal of Cognitive Engineering and Decision Making* 2(2): 140–160.

Parasuraman, R., and C. D. Wickens. 2008. Humans: Still vital after all these years of automation. *Human Factors* 50: 511–520.

Parasuraman, R., and G. F. Wilson. 2008. Putting the brain to work: Neuroergonomics past, present and future. *Human Factors* 50: 468–474.

Pettersson, P. O., and P. Doherty. 2004. Probabilistic roadmap based planning for an autonomous unmanned helicopter. *Sensors*: 1–6.

Proud, R. W. (2005). *The Function-Specific Level of Autonomy and Automation.* Johnson Space Ceneter, http://research.jsc.nasa.gov/PDF/Eng-16.pdf (accessed June 9, 2010).

Proud, R. W., and J. J. Hart (2005). *FLOAAT: A Tool for Determining Levels of Autonomy and Automation, Applied to Human-Rated Space Systems.* AIAA infotech@Aerospace 2005. Arlington, VA: American Institute of Aeronautics and Astronautics.

Reid, G. B., C. A. Shingledecker, and T. Eggemeier. 1981. Application of conjoint measurement to workload scale development. In *Proceedings of the Human Factors 25th Annual Meeting,* 522–525. Santa Monica, CA: Human Factors Society.

Reeves, B., and C. Nass. 1996. *The Media Equation: How People Treat Computers, Television and New Media Like Real People and Places.* Cambridge, MA: Cambridge University Press.

Rice, S. 2009. Examining single- and multiple-process theories of trust in automation. *Journal of General Psychology* 136(6): 303–319.

Riley, V. A. 2000. Developing a pilot-centered autoflight interface. In *Proceedings of the World Aviation Congress and Exposition,* 241–245. Warrendale, PA: SAE International.

Rouse, W. B. 1988. Adaptive aiding for human/computer control. *Human Factors* 30: 431–443.

Russell, C. A., and G. F. Wilson. 1998. Air traffic controller functional state classification using neural networks. In *Proceedings of the Artificial Neural Networks in Engineering (ANNIE '98) Conference,* 649–654. New York: American Society of Mechanical Engineers.

Russell, C. A., G. E. Wilson, and C. T. Monett. 1996. Mental workload classification using backpropagation neural network. In *Intelligent engineering systems through artificial neural networks,* by C. H. Dagli, M. Akay, C. L. P. Chen, B. R. Fernandez, and J. Ghosh, 685–690. New York: American Society of Mechanical Engineers.

Sarter, N., D. D. Woods, and C. E. Billings. 1997. Automation surprises. In *Handbook of Human Factors and Ergonomics* (2nd ed.), ed. G. Salvendy, 1926–1943. New York: Wiley.

Scallen, S., P. A. Hancock, and J. A. Duley. 1995. Pilot performance and preference for short cycles of automation in adaptive function allocation. *Applied Ergonomics* 26: 397–403.

Scerbo, M. 1996.Theoretical perspectives on adaptive automation. In *Automation and Human Performance: Theory and Applications,* by R. Parasuraman and M. Mouloua, 37–63. Mahwah, NJ: Erlbaum.

Scerbo, M. 2007. Adaptive automation. In *Neuroergonomics: The Brain at Work,* by R. Parasuraman and M. Rizzo, 238–252. New York: Oxford University Press.

Sheridan, T., and R. Parasuraman. 2006. Human-automation interaction. In *Reviews of Human Factors and Ergonomics,* ed. R. S. Nickerson, Vol. 1, 89–129. Santa Monica, CA: Human Factors and Ergonomics Society.

Sheridan, T. B., and R. W. Simpson. 1979. *Toward the Definition and Measurement of the Mental Workload of Transport Pilots.* Cambridge, MA: Massachusetts Institute of Technology.

Sheridan, T. B., and W. L. Verplank. 1979. *Human and Computer Control of Undersea Teleoperators.* Cambridge, MA: Massachusetts Institute of Technology.

Squire, P., G. Trafton, and R. Parasuraman. 2006. *Human Control of Multiple Unmanned Vehicles: Effects of Interface Type on Execution and Task Switching Times.* HRI 06. Salt Lake City, UT.

Sterman, M. B., and C. A. Mann. 1995. Concepts and applications of EEG analysis in aviation performance evaluation. *Biological Psychology* 40: 115–130.

Verwey, W. B., and H. A. Veltman. 1996. Detecting short periods of elevated workload: A comparison of nine workload assessment techniques. *Journal of Experimental Psychology: Applied* 2: 270–285.

Wickens, C. D. 1979. Measures of workload, stress and secondary tasks. In *Mental Workload: Its Theory and Measurement*, by N. Moray, 79–99. New York: Plenum.

Wickens, C. D. 1984. Processing resources in attention. In *Varieties of Attention*, ed. R. Parasuraman and D. R. Davies, 63–102. Orlando, FL: Academic Press.

Wickens, C. D. 2008a. Functional allocation and the degree of automation. Presentation at the Rocky Mountain Chapter of HFES. http://Function_allocation_and_the_degree_of_automation_C_Wickens_16_Oct_2008.pdf (accessed June 6, 2010).

Wickens, C. D. 2008b. Situation awareness: Review of Mica Endsley's 1995 articles on SA theory and measurement. *Human Factors* 50: 397–403.

Wickens, C. D., and S. R. Dixon. 2007. The benefits of imperfect diagnostic automation: A synthesis of diagnostic atuomation in simulated UAV flights: An attentional visual scanning analysis. In *Proceedings of the 13th International Symposium on Aviation Psychology,* 818–823. Dayton, OH: Wright-Patterson Air Force Base.

Wickens, C. D., and J. G. Hollands. 2000. *Engineering Psychology and Human Performance.* Upper Saddle River, NJ: Prentice Hall.

Wicks, A. C., S. L. Berman, and T. M. Jones. 1999. The structure of optimal trust: Moral and strategic implications. *Academy of Management Review* 24: 99–116.

Wilson, G. F. 2001. In-flight psychophysiological monitoring. In *Progress in Ambulatory Monitoring*, ed. F. Fahrenberg and M. Myrtek, 435–454. Seattle, WA: Hogrefe and Huber.

Wilson, G. F. 2002. Psychophysiological rest methods and procedures. In *Handbook of Human Factors Testing and Evaluation*, ed. S. G. Charlton and G. O'Brien, 157–180. Mahwah, NJ: Erlbaum.

Wilson, G. F., and F. T. Eggemeier. 1991. Physiological measures of workload in multi-task environments. In *Multiple-task performance*, ed. D. Damos, 329–360. London: Taylor & Francis.

Wilson, G. F., and F. Fisher. 1995. Cognitive task classification based upon topographic EEG data. *Biological Psychology* 40: 239–250.

Wilson, G. F., and C. A. Russell. 2003a Operator functional state classification using multiple psychophysiological features in a air traffic control task. *Human Factors* 45: 381–389.

Wilson, G. F., and C. A. Russell. 2003b. Real-time assessment of mental workload using psychophysiological measures and artificial neural networks. *Human Factors* 45: 635–643.

Wilson, G. F., and C. A. Russell. 2007. Performance enhancement in an uninhabited air vehicle task using psychophysiologically determined adaptive aiding. *Human Factors* 49: 1005–1018.

Woods, D. 1996. Decomposing automation: Apparent simplicity, real complexity. In *Automation and Human Performance: Theory and Application*, ed. R. Parasuraman and M. Mouloua, 3–17. Mahwah, NJ: Erlbaum.

8 Safety Assessments

Eric J. Shappee

CONTENTS

8.1 INTRODUCTION

For years the aviation field has been rapidly advancing in technology. With all the change, aviation organizations and manufacturers have found themselves faced with new safety issues and ever-changing safety requirements. The unmanned aircraft system (UAS) field is no different. In fact, safety in this arena is more of a concern. With no onboard pilot, complex operating systems, and ever-changing avionics as well as continuous software updates, safety appears to be one of the major hurdles for integrating UAS into the National Airspace System (NAS).

This chapter will examine several safety tools and techniques such as hazard analysis and its various forms. It will also cover the risk assessment process and provide some guidance on developing a risk assessment tool. And finally, it will look at safety evaluations and provide some thoughts on UAS accident investigation considerations.

8.2 HAZARD ANALYSIS

The hazard analysis can take several forms. In this section we will look at several common types of hazard analysis. The purposes and function of the hazard analysis are all predicated on what stage of the operation for which you are applying it.

8.2.1 Purpose

Hazard analyses are common tools found in the system safety arena. Generally these tools are used throughout various stages of a product life cycle. In his book *System Safety for the 21st Century*, Richard Stephans identifies the various stages of a product life cycle. These stages or phases are concept, design, production, operations, and disposal. Although in UAS operations we are not looking specifically at the development of a product throughout its life cycle, we are, however, looking at its operational phase. We can subdivide the UAS operational phase into several general stages: planning, staging, launch, flight, and recovery. Applying the appropriate hazard analysis tool within each stage will allow for early identification and ultimately early resolution of safety issues.

8.2.2 Preliminary Hazard List

The preliminary hazard list (PHL) is just what it sounds like, a list. Simply put, it is a brainstorming tool used to identify initial safety issues early in the UAS operation. To get the most out of the PHL, you need to have a variety of input from the people familiar with each stage of the UAS operation and its phases. Figure 8.1 is an example of a PHL that can be used to aid in the process.

To use the PHL tool, we first need to have an in-depth understanding of the stage we are going to evaluate. At the top of the form select the stage (planning, staging, launching, flight, or recovery) to be evaluated. Doing this helps to keep all the various sheets from team members organized in the proper categories for ease of review. The next step is to list a tracking number (1, 2, 3, etc.) and potential hazards we see in the selected stage. For example, in the staging phase you may want to list items such as nearby terrain features (trees, power lines and poles, and antennas). After listing the hazards, we need to determine the probability and severity of the hazard. In the probability column we can enter frequent, probable, occasional, remote, or improbable. These probability levels are listed and defined in MIL-STD-882D/E and in the Appendix of this text.

The next column is severity. In this column you can use the categories of catastrophic, critical, marginal, or negligible. Like the probability levels, the severity categories and definitions are also listed in MIL-STD-882D/E and in the Appendix.

PRELIMINARY HAZARD LIST/ANALYSIS (PHL/A)							
DATE: ______		PREPARED BY: ______				Page ____ OF ____	
Operational Stage:	☐ Planning	☐ Staging	☐ Launch	☐ Flight		☐ Recovery	
TRACK#	HAZARD	PROBABILITY	SEVERITY	RL	MITIGATING ACTION	RRL	NOTES
RL = Risk Level, RRL = Residual Risk Level				Probability, Severity, and Risk Levels defined in MIL-STD-882D/E			

FIGURE 8.1 Preliminary hazard list/analysis (PHL/A).

The last column of the hazard list is the RL or risk level. This is the point where we establish an initial risk level value based on the probability and severity that we have identified. For instance, if we determined that launching the UAS at a field that has trees nearby would have a probability of impacting a tree to be remote and the severity to be critical, then using the risk matrix in MIL-STD-882D/E we can determine that the risk RL for that hazard is a 10. Note that the higher the number, the lower the risk. If you decide to develop your own preliminary hazard list or analysis (PHL/A) be careful, not all risk matrices are alike; some will be organized to have the lower number signify a lower risk.

8.2.3 Preliminary Hazard Analysis

Once the initial risk levels have been identified, we now need to move into the analysis phase by looking at ways to mitigate the listed hazard. This is fairly simple; here is where we ask what can be done to reduce or eliminate the hazard. When looking at mitigation we need to look at it in terms of probability and severity. Concerning probability, we want to determine ways to eliminate or reduce the possibility of occurrence, or better known as exposure. Let us say that we have determined that the field in which we want to operate out of has trees at the approach and departure ends of the runway. In the mitigating action column we can list several solutions. The first can be relocating to another field with no trees; the second possible action could be to remove the trees; and the third, and probably the most reasonable, would establish or modify the launch and recovery procedures.

The next column is the RRL or residual risk level column. This time we are asking, will we lower the risk by implementing these mitigating actions. Just like when we determine the RL, we have to consider the probability and severity. You may find that one or both (probability, severity) may have changed. Changing any one of these factors can lower or increase the risk level. Obviously if we increase the risk level we do not want to implement that particular mitigating action.

The last column is labeled notes. This is fairly self-explanatory. If we have any special concerns or instructions needed for the implementation of the mitigating actions, we will want to list them in this column in some detail. As we complete the PHL/A worksheet we must keep in mind that it is used as an initial hazard identification tool. Once the UAS activity is underway, an operational hazard analysis should be performed to evaluate the hazards after the mitigating action have been applied. Hazard analysis tools like the PHL/A are extremely useful when assessing the hazards of the UAS operational cycle. The main purpose for using the hazard analysis tool is to provide the user with a systematic approach to identifying, analyzing, and mitigating hazards early in the operation.

8.2.4 Operational Hazard Review and Analysis

Just as the PHL/A tool is used to identify initial safety issues early in the UAS operation, the operational hazard review and analysis (OHR&A) is used to identify and evaluate hazards throughout the entire operation and its stages (planning, staging, launching, flight, and recovery). This is a crucial part of the ongoing and continuous evaluation of hazards and provides the feedback necessary to determine that the mitigating actions employed have worked as expected.

Obviously we would want to continue monitoring the hazards we listed on the PHL/A, but there may be other hazards that appear during the UAS operation or activity that were not foreseen. Items that you should always consider with the OHR&A are in the area of human factors. These items are human interface with the equipment and operating systems as well as crew resource management (CRM). This can get complicated quickly, depending on the number of crew members and their specific tasks. Both human factors and CRM will be covered in more depth in later chapters but human factors issues and CRM must be continuously monitored.

The use of the OHR&A tool is very similar to the PHL/A (Figure 8.2). The main difference is the action review column. In this column we want to list if the identified mitigating actions implemented from the PHL/A were adequate. If the actions were not adequate and the hazard has not changed, then list the hazard again. If the actions have modified the hazard, then list the modified one. At this point the rest of the OHR&A tool works like the PHL/A. To aid in keeping things organized I suggest the use of separate worksheets when it come to hazard review and the evaluation of new operational hazards. I also suggest the tracking numbers on the OHR&A sheet correspond with the ones listed on the PHL/A. Doing this will aid in keeping all the safety analysis and review information organized. Just as before, the probability levels, severity categories, and risk matrix are listed in MIL-STD-882D/E and in the Appendix.

OPERATIONAL HAZARD REVIEW & ANALYSIS (OHR&A)							
DATE: ________		PREPARED BY: ____________				Page ____ OF ____	
Operational Stage:	☐ Planning	☐ Staging		☐ Launch	☐ Flight	☐ Recovery	
TRACK#	ACTION REVIEW	PROBABILITY	SEVERITY	RL	MITIGATING ACTION	RRL	NOTES
RL = Risk Level, RRL = Residual Risk Level				Probability, Severity, and Risk Levels defined in MIL-STD-882D/E			

FIGURE 8.2 Operational hazard review and analysis (OHR&A).

8.2.5 Change Analysis

The change analysis serves a crucial role in the ongoing review and analysis of safety. What the change analysis allows you to do is review and examine any changes that have been made to the operation. For example, if we have a UAS system software change such as an upgrade for the UAS computers or operating systems, we will want to make an assessment of the changes and evaluate how these changes affect the overall operation. Another example would be a procedural change; you may have modified the launch procedure to get the vehicle in the air faster. This modification would also warrant an assessment of the changes made. To assess the change, use the OHR&A worksheet. List all hazards associated with the changes in the action review column and run the worksheet as you would an OHR&A.

8.3 RISK ASSESSMENT

According to Maguire (2006), "public perception of risk is the key to safety" (p. 47). I would dare to take it a step further and state that in the unmanned aircraft world, the public perception of risk is the key to airspace integration and acceptance. How we approach and manage that risk is critical. One type of tool that has been used by the military, airlines, and some flight training schools is a basic risk assessment matrix. The risk assessment tool in Figure 8.3 is a derivative of one that was

sUAS RISK ASSESSMENT

2/20/10 Kansas State University at Salina

UAS Crew/Station: ________/____ ________/____

________/____ ________/____

Mission Type	SUPPORT 1	TRAINING 2	PAYLOAD CHECK 3	EXPERIMENTAL 4	
Hardware Changes	NO 1			YES 4	
Software Changes	NO 1			YES 4	
Airspace of Operation	Special Use 1	Class C 2	Class C 3	Class E, G 4	
Has PIC Flown This Type Aircraft	YES 1			NO 4	
Flight Condition	DAY 1			NIGHT 4	
Visibility	≥10 MILES 1	6–9 MILES 2	3–5 MILES 3	<3 MILES 4	
Ceiling in Feet AGL	≥10,000 1	3000–4900 2	1000–2900 3	<1000 4	
Surface Winds		0–10 KTS 2	11–15 KTS 3	>15 KTS 4	
Forecast Winds		0–10 KTS 2	11–15 KTS 3	>15 KTS 4	
Weather Deteriorating	NO 1			YES 4	
Mission Altitude in Feet AGL		<1000 2	1000–2900 3	≥3000 4	
Are All Crew Members Current	YES 1		NO 3 ---►	CURRENCY FLIGHT REQUIRED	
Other Range/ Airspace Activity	NO 1			YES 4	
Established Lost Link Procedures	YES 1			NO NO FLIGHT	
Observation Type	Line of Sight & Chase 1		Chase Only 3	Line of Sight & Only 4	
UAS Grouping	GROUP I 1	GROUP II 2	GROUP III 3	GROUP IV 4	
Total					

RISK LEVEL			
20–30 LOW	31–40 MEDIUM	41–50 SERIOUS	51–64 HIGH

Aircraft Number: ____________ Aircraft Type: ____________

Flight Released By: ____________ Date: ________ Time: ________

FIGURE 8.3 Small UAS risk assessment.

developed for a flight-training program. Risk assessment can best be defined as the evaluation of common operational hazards in terms of severity and probability.

8.3.1 Purpose

The risk assessment tool serves two purposes. The first is it provides the UAS/RPA operator with a quick look at the operation before committing to the flight activity (a go/no-go decision). The second is that it allows safety and management of real-time information needed to continually monitor the overall safety of the operation. This tool should be completed by the UAS/RPA operator before each flight and briefed to the crew. The briefing should consist of at least a review of the risks, hazards, and any concerns associated with the activity. This tool is meant to be an aid in the decision-making process and should not be the only means used in making the go/no-go decision.

8.3.2 Development

The risk assessment tool in Figure 8.3 was designed for small UAS operations. As stated earlier, the risk assessment tool is meant to be an aid in the decision-making process. When considering developing a risk assessment tool you will want to tailor it to your specific operation. To get started, assemble those directly involved with the operation and discuss the operational factors such as weather, crew rest, and airspace. Also, you should consider items listed on the PHL/A that would change per flight cycle.

Once you have developed the list, the next step is to identify how each factor can change in terms of probability and severity. At this point you will need to make a decision on whether to use a numeric ranking scale. If you choose not to, that is fine; the only caution I give, however, is that you may not have an easy way to identify the overall total risk level (low, moderate, serious, and high). My recommendation would be to add some type of ranking system. The one in Figure 8.3 is a numeric system with a total value scale listed at the bottom of the sheet. Attached to each total value scale is a total risk level category (low, moderate, serious, and high). These categories along with an example risk index are listed in MIL-STD-882D/E and in the Appendix. The numeric scale makes computer tracking and monitoring of overall operational risk easier. The overall risk category aids in briefing team members and gives them a meaningful risk level for the operation.

The last item that needs to be mentioned and considered under development is reviewing and updating the risk assessment tool. Periodically review the effectiveness of the tool and make changes as necessary. You may find that some of the factors identified have changed. This could be due to a platform change or the operational factor or hazard was eliminated. Also review the OHR&A and change analysis to determine if any of the new hazards identified need to be considered.

8.3.3 Use

To use the risk assessment form in Figure 8.3, you will want to start by listing the crew members and their position or station. Next, move to the matrix and start with the left-hand column where you see the first operational factor, mission type. From this point move right until you reach the type of mission. The choices listed are support, which covers a broad range of activities such as disaster response; training, an example would be a new UAS operator; payload check, which covers upgrades to payload or new payloads; and experimental, which would be classified as a new vehicle or type or UAS operation.

Looking at the first row, if your mission type is training, the associated risk number would be 2 and you place a 2 in the far-right column. If your mission type is experimental, place a 4 in the far-right column. Continue down the operational factors list in the left column and move right to the associated risk level that fits your flight, and place that number in the far-right column. As you can see, the farther right you go, the greater the associated risk level. Once you have determined the risk levels for each operational factor, add the numbers in the far-right column to determine your total risk value.

Once the total risk value has been calculated, find which range your value falls within. For instance, if your total value is 26, the risk is low. The risk levels of low, medium, serious, and high are derived from MIL-STD-882D/E. Below the risk levels you will find spaces for aircraft number, aircraft type, flight released by, date, and time. All are self-explanatory except for "flight released by." This space should be reserved for someone with management authority such as the chief pilot, mission director, and so forth. The idea behind this is to have management review each evaluated operational factor as well as the total overall risk value and sign for risk acceptance. Remember that this is just a tool to help assess the risk and safety of the operation. This tool should not be the only means of determining a mission go or no-go.

Looking back at the matrix section, a few of the operational factors listed in the left column warrant further explanation. Hardware changes are items such as wing sets and engines. Items like operating system updates or new versions of software would fall under software changes. Under the operational factor of airspace of operation, you will find "special use." A good example of "special use" would be restricted areas or an area with temporary flight restrictions (TFR). Also in the same row you will find Class C, D, E, and G airspaces. As the involvement of air traffic control (ATC) is decreased you will notice that the risk level increases. Currently, the two predominant airspaces for civil UAS operations are restricted areas, E and G, provided you are operating under some type of authorization or waiver such as a certificate of authorization (COA). When it comes to other range/airspace activities you have a choice of yes or no. The idea behind this operational factor is that if you have other aircraft in the vicinity/airspace or restricted area they could constitute a hazard and should be considered. The last item is the UAS grouping. Detailed information on these groups can be found on the Federal Aviation Administration (FAA) Web site (www.faa.gov). In general, these groupings address a variety of items. Some of the items addressed are weight limits, speed limitations, and altitude restrictions.

8.4 SAFETY EVALUATION

A major key to integrating UAS into the national airspace is its safety evaluation. The FAA recognizes that UAS will need to meet an acceptable level of risk. To do this can be very challenging. This section will examine several ways to aid in the evaluation process of operational safety. Items that will be discussed will be risk assessments, flight test cards, and airworthiness.

8.4.1 Risk Assessment

As stated earlier, the purpose of risk assessment is twofold. First, it provides the UAS operator with a quick look at the operation before committing to the flight activity (a go/no-go decision). (Note: A risk assessment should be completed before every flight activity.) Second, it allows safety and management the means to review the operational risks and continually monitor the overall safety. It is this review and continuous monitoring along with the completed risk assessment tools that provides the needed data to show an acceptable level of risk with the flight operation.

8.4.2 Flight Test Cards

Another key element to the safety evaluation is the flight test card. A flight test card is a set of tasks or functions that the UAS vehicle and/or ground station must be able to perform. These test cards are usually performed in some type of special use airspace such as a restricted area where a FAA authorization or waiver such as a COA is not required. After all, the whole purpose of the safety evaluation is to develop good safe practices and gain the needed safety data for FAA authorization to fly in the national airspace.

The test card shown in Figure 8.4 is the final flight test card completed before the university will endorse the airworthiness certification. When developing test

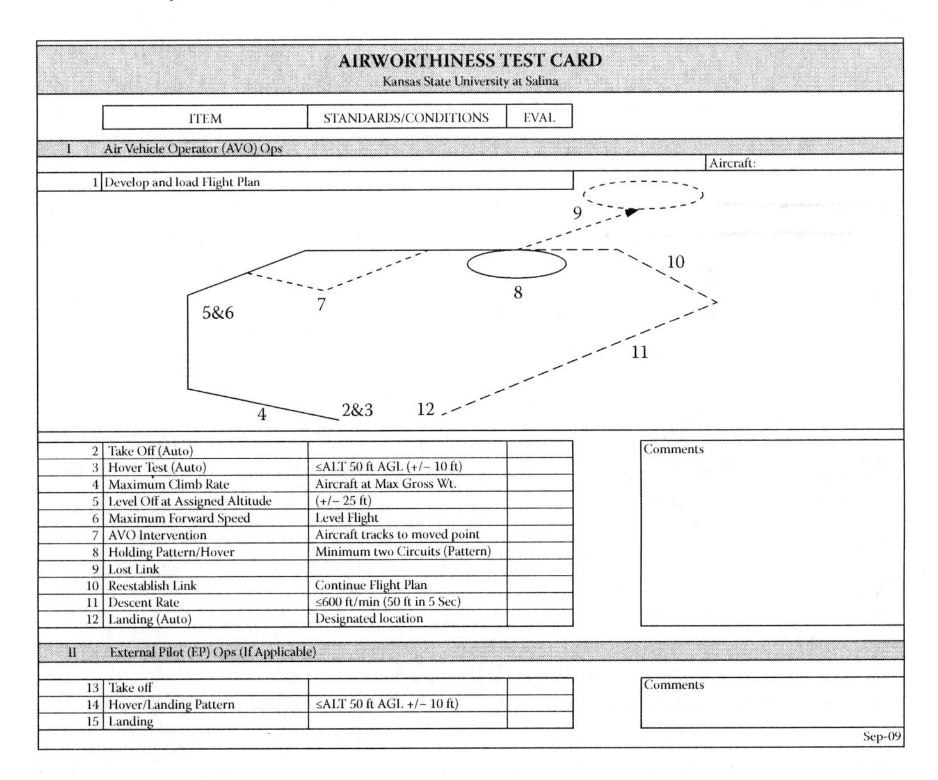

AIRWORTHINESS TEST CARD

Kansas State University at Salina

	ITEM	STANDARDS/CONDITIONS	EVAL
I	Air Vehicle Operator (AVO) Ops		
1	Develop and load Flight Plan		

Aircraft:

	ITEM	STANDARDS/CONDITIONS	EVAL
2	Take Off (Auto)		
3	Hover Test (Auto)	≤ALT 50 ft AGL (+/− 10 ft)	
4	Maximum Climb Rate	Aircraft at Max Gross Wt.	
5	Level Off at Assigned Altitude	(+/− 25 ft)	
6	Maximum Forward Speed	Level Flight	
7	AVO Intervention	Aircraft tracks to moved point	
8	Holding Pattern/Hover	Minimum two Circuits (Pattern)	
9	Lost Link		
10	Reestablish Link	Continue Flight Plan	
11	Descent Rate	≤600 ft/min (50 ft in 5 Sec)	
12	Landing (Auto)	Designated location	

Comments

	ITEM	STANDARDS/CONDITIONS	EVAL
II	External Pilot (EP) Ops (If Applicable)		
13	Take off		
14	Hover/Landing Pattern	≤ALT 50 ft AGL +/− 10 ft)	
15	Landing		

Comments

Sep-09

FIGURE 8.4 Airworthiness test card.

cards such as the airworthiness, auto land, or payload specific tests, you need to have an understanding of the equipment being used and its limitations. You also need to be familiar with the FAA requirements for UAS operations. If these test cards are developed properly, they will be a great asset along with risk assessment. These two tools can go a long way in providing operational safety data for airworthiness certification.

8.4.3 Airworthiness Certification

According to the FAA, public institutions have the option to self-certify airworthiness. Doing this should involve a few more steps than just saying "everything looks good; we are airworthy; let's go fly." Using the tools discussed in this chapter, like the risk assessments and flight test cards, will be very valuable in this process. However, we still need to consider operator and crew qualifications, air vehicle reliability, ground station reliability, and program/software capabilities before we even fly. When it comes to the operator and crew, we need to review their qualifications such as pilot certifications, experience, and competency using the system. As for air vehicle reliability, we need to consider structural integrity, power plant (engine) reliability, and aerodynamics and performance. When considering ground station reliability, we are asking, how reliable is the equipment. Is there a back up or contingency plan for equipment failures? The last consideration listed is program and software capabilities. Are the programs user friendly to minimize human factors issues? How reliable is the software and programs that are being used? Are there any backup systems? Are there any frequency issues/conflicts? As you can see, the questions can be virtually endless. Just as critical as the items that are being evaluated is having some way of documenting this information, be it an application, checklist, or a combination there of. These are but just a few basic areas that need to be examined along with some questions that would need answers in order to fly the air vehicle for further safety evaluation.

8.5 ACCIDENT INVESTIGATION CONSIDERATIONS

One subject about which there is very little information available is UAS accident investigation. Although many of the traditionally used tools and techniques used in the manned aircraft type investigations will work in unmanned aircraft, there are some unique differences.

8.5.1 Software and Hardware

Most of us are fairly familiar with how to operate the programs that are installed on our home computers or laptops. But, are we really familiar with the software? Do we really know all the features of the programs that we use? Unless you are a computer guru, the answer is probably no. When it comes to software, many safety professionals, especially the system safety folks, know that software, if not compatible with other software or operating systems can cause serious problems.

Stephans (2004, p. 53) states, "A software specification error, design flaw, or the lack of generic safety critical requirement can contribute to or cause a system failure or erroneous human decision." When it comes to investigating UAS accidents you will want to take a close look at software. To do this you will probably need someone who is very familiar with the specific operating system such as the software engineer or a programmer.

Like software, hardware is a critical area in accident investigation. The hardware components can be divided into two categories. First is the hardware configuration. Here we would want to ask if all components have been connected and physically checked for proper configuration. Examples would be transmitters, backup power supplies, and antennas. The second category, and the one that could be the most problematic, is the interface between the hardware and software. Here is where you need to ask yourself if the hardware and software are compatible? Again you will need someone who is very familiar with the specific operating software and components.

As the person responsible for conducting UAS accident investigations it would be to your advantage to have someone with these special skills in this area on your team or party. Another benefit to having members who are very familiar with the software and hardware of the system is that most of the UAS systems record flight and operational data. With the expertise of these individuals, they should be able to extract and interpolate the data from the flight. They can also be beneficial in aiding you in the simulation and reconstruction of the entire flight. This type of information that is extracted is similar to the information that is retrieved from a flight data recorder (FDR).

8.5.2 Human Factors

Although human factors are covered in another chapter, they warrant a special note in this section. As time goes on, you will see more studies concerning human factors and UAS operations. In this section, I offer some areas that you may want to consider when investigating a UAS incident or accident. The first is crew coordination. Unlike the airlines with two or three cockpit crew members, with UAS/RPA operations you can find significantly more crew than just the pilot and copilot, who in many cases is the payload operator. If your system is not equipped with an auto land or takeoff system, you will have added an external pilot with a remote control (RC). In most operations you will need an observer or chase plane with pilot. This adds another level of complexity for crew coordination.

System complexity or user-friendliness is another issue that you will want to consider. Many of the operating systems and associated software were designed by computer engineers with no aviation experience. What is simple for them may not be simple or even flow on a checklist for the UAS operator. A system not designed with human factors in mind could see a high increase in operator error. This error could occur any where from mission planning and programming to a situation where time is critical, such as quick flight plan change for collision avoidance.

INVESTIGATION ROSTER

IIC____________________ Case #__________ Start Date__________

	Section Chair					
Operations						
Pilot/Maintenance						
Safety Assessments						
Human Factors						
Software/Hardware						
ATC (If Applicable)						
Weather/Airfield						
Structures/Perf						
Witnesses						
Other						

FIGURE 8.5 Investigation roster.

8.5.3 Suggestions

If you get tasked with investigating a UAS accident, I have a couple suggestions. First, do not try to tackle the investigation all on your own. You are going to need experts familiar with the field to help you get and analyze the information. Second, have a plan to get organized; know what the major areas are for the investigation. One tool that I use is an investigation roster shown in Figure 8.5. This roster lists the major areas and provides space for assigning team members to specific tasks.

8.6 CONCLUSION AND RECOMMENDATIONS

The information provided in this chapter has provided a good starting point for UAS safety and safety evaluations. The tools discussed in this chapter are tools that I have developed over the past several years and have used during the evaluation of UASs that I have flown and for which I have obtained COAs. As the world of remotely piloted aircraft grows, so will the need for safety. For those of you who are interested in broadening your knowledge of safety and jumping into this field with both feet, I would like to offer a couple of suggestions. First, take some safety courses. Take courses in the area of safety management, system safety, and safety management systems. I have found these courses to be invaluable when developing safety tools and evaluating safety of operations. Second, look at the reference sections in this book and spend some time online or at the library reviewing some of the references listed from this chapter.

DISCUSSION QUESTIONS

8.1 List and discuss each of the UAS operational phases.
8.2 Define probability and severity.
8.3 Discuss the difference between the PHL/A and the OHR&A.
8.4 What are the two purposes of the risk assessment?
8.5 What is the purpose of the safety evaluation?
8.6 Discuss some of the differences between manned-type accident investigation and unmanned.

REFERENCES

Federal Aviation Administration. 2009. Small Unmanned Aircraft Systems Aviation Rulemaking Committee: Comprehensive Set of Recommendations for sUAS Regulatory Development.
Department of Defense. 2000. MIL-STD-882D—Standard Practice for System Safety.
Maguire, R. 2006. *Safety Cases and Safety Reports*. Burlington, VT: Ashgate.
Shappee, E. 2006, March. Grading the go. *Mentor* 8: 12.
Stephans, R. 2004. *System Safety for the 21st Century*. Hoboken, NJ: Wiley-IEEE.

9 Detect, Sense, and Avoid

Lisa Jo Elliott, Jeremy D. Schwark, and Matthew J. Rambert

CONTENTS

9.1 INTRODUCTION

9.1.1 Detect, See, and Avoid: Manned Aircraft

The United States Federal Aviation Administration (FAA) has long relied upon the eyesight of a human pilot as the primary method to avoid midair collisions even when transponders or radar systems are present. Lacking a human pilot, unmanned aircraft systems (UASs) do not have the advantage of this onboard sense-and-avoid safety feature. An increasing number of military, civilian, and commercial applications for UASs may lead to an increasingly crowded airspace.

The FAA, tasked with regulation of the nation's airspace, is concerned with UAS resolution of the detect, see, and avoid (DSA) problem. According to (technical reports) 14 CFR 91.113 and RTCA DO-304 (Guidance Material and Considerations for Unmanned Aircraft Systems), when UASs share the same airspace with manned air vehicles, it will be necessary that automated sense-and-avoid systems provide a level of safety equaling or exceeding that of manned aircraft. In July 2004, an endeavor to set the standards for this "equivalent level of safety" was attempted when the American Society for Testing and Materials Standards subcommittee released Document F2411-04 (since amended to F2411-04e1), Standard Specification for Design and Performance of an Airborne Sense-and-Avoid System. Since its release, this document has served as a guideline for developers and researchers working on UASs.

However, this document does not cover operation of UASs in the National Airspace System (NAS). Until recently, the Federal Aviation Regulations (FARs) outlined regulations for the operation of moored balloons, kits, unmanned rockets, and unmanned free balloons (14 CFR Part 101), but not for UASs. To address this, the FAA recently issued a memorandum titled "ASF-400 UAS Policy 05-01," dated September 16, 2005, which updates previous guidance. The FAA uses this latest policy to determine if a UAS may operate in the NAS and acknowledges the problem of UAS to comply with the duty to "see and avoid" other aircraft. Operations in the NAS that fall short of this mandate will not be authorized, including UAS operations.

9.1.2 Pilot's See-and-Avoid Role: Manned Aircraft

A pilot has several responsibilities defined in the FARs and Aeronautic Information Manual, one of which is detect, see, and avoid. In a UAS, this must be accomplished by a technology solution or human observer external to the UAS (RTCA DO-304). Seeing and avoiding other aircraft is a difficult task for the general aviation pilot (FAA, 2006). A general aviation pilot must be able to monitor instruments, tune radios and communicate, tune navigational equipment, read maps, navigate, and fly the aircraft. Many pilots fly with portable global positioning system (GPS) devices, which improve situational awareness but also turn the pilot's attention from outside the aircraft to inside the aircraft. Most pilots recognize their limitations at seeing and avoiding aircraft, so they rely on established procedures (e.g., landing approach patterns) to ensure they maintain separation from other aircraft.

9.1.3 Detect, See, and Avoid: UASs

As previously mentioned, a certificate of authorization (COA) is issued by the FAA to ensure an equivalent level of safety to that of manned aircraft. To expand, the FAA states that a UAS (having no onboard pilot) requires special provisions outside of restricted, prohibited, or warning areas. This provision can be fulfilled by using visual observers, either ground based or airborne. Observers are to fulfill the same DSA duties as an onboard pilot: to see traffic that may be in conflict, evaluate flight paths, determine traffic right-of-way, and maneuver to avoid the traffic.

Detecting and sensing something in the NAS is defined as determining the presence of something, not necessarily identified, in your airspace. This is both a sensing and a judgment task. The pilot or visual observer must determine if an object is indeed present, sensing. Then, the observer must determine if the detected object is or is not a threat or target. A decision must be made to enact procedures as a result of a positive judgment. These three activities define the steps an autonomous system must fulfill to mimic the performance of a human observer during DSA.

9.2 SIGNAL DETECTION APPROACH FOR DETECT, SEE, AND AVOID

Signal detection theory (SDT) long has been used as a means of characterizing the performance of humans during target identification. SDT began with the development of radar and communications equipment in the early part of the 20th century. The model describes the human sensory and perceptual system during detection of ambiguous stimuli. The book by Green and Swets (1966), *Signal Detection Theory and Psychophysics,* provides a more complete explanation of the model and the historical underpinnings.

SDT assumes that the decision maker is actively engaged and interacting with a constantly changing environment. For example, a human observer has a goal to achieve (directing flight). As the goal is pursued, many decisions must be made. Among those decisions is DSA. As the person perceives an object, the perception is influenced by many factors. There may be many other objects present at the same time, internally and externally. Generally, these nontarget objects are referred to as noise. Externally present objects that interfere with detection are said to be external noise (e.g., fog or lighting). Internally present objects that interfere with detection are said to be internal noise (e.g., fatigue or drug use). During DSA, the observer must first decide if the object is present. Table 9.1 describes the four outcomes of this decision.

9.2.1 Response Bias and Response Criterion

The person's response decision on whether to accept or reject the perceptual information is based on a bias. Persons with similar perceptual information may make different decisions depending on their bias (ß). The bias can vary significantly depending on the situation and the consequences of a decision. If the perceptual input is greater than the ß, the person will adopt a positive identification (target

TABLE 9.1
Four Outcomes of a Signal Detection Decision

		Signal	
		Present	**Absent**
Decision	Visible	Hit (e.g., the observer reports that the evening star is visible in the sky and actually it is visible in the sky)	False alarm (e.g., the observer reports the evening star is visible but actually it is not visible in the sky)
	Absent	Miss (e.g., the observer reports that the evening star is not visible in the sky and actually it is visible in the sky)	Correct rejection (e.g., the observer reports that the evening star is not visible in the sky and actually it is not visible in the sky)

is present). If the perceptual input is less than ß, the person will adopt a negative identification (target is absent).

The bias can be modeled with distribution graphs as shown in Figure 9.1. A neutral bias has a ß level at or around 1.0. This might occur in situations where the consequences for a negative or a positive identification are equally matched.

A conservative bias has a ß level greater than 1.0. In this instance, the person is more likely to give a negative identification (no object detected) (Figure 9.2). This might occur in situations where the consequences for a false alarm (identifying a target when none is present) are greater than the rewards of a hit (identifying a target when one is present).

A liberal bias has a ß level less than 1.0. In this instance, the person is more likely to give a positive identification (an object is detected) (Figure 9.3). This might occur in situations where the consequences of a miss (failing to identify a target when there is one) are greater than the consequences for a false alarm.

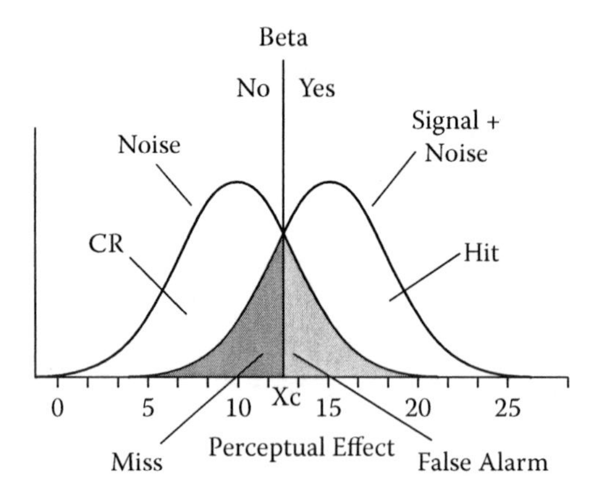

FIGURE 9.1 Neutral bias, equal chance of having either a positive or a negative response.

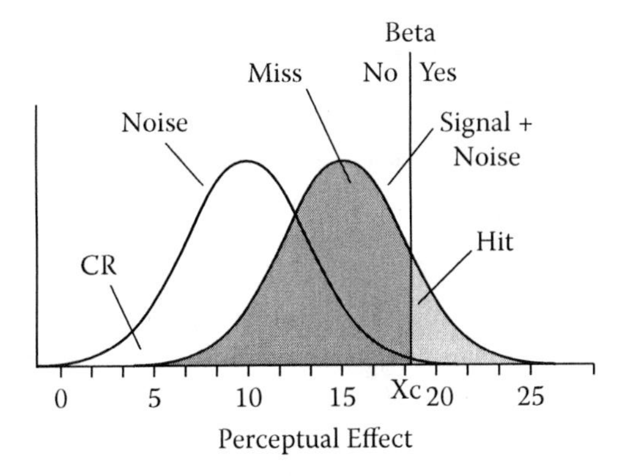

FIGURE 9.2 Conservative bias, greater chance of a negative response.

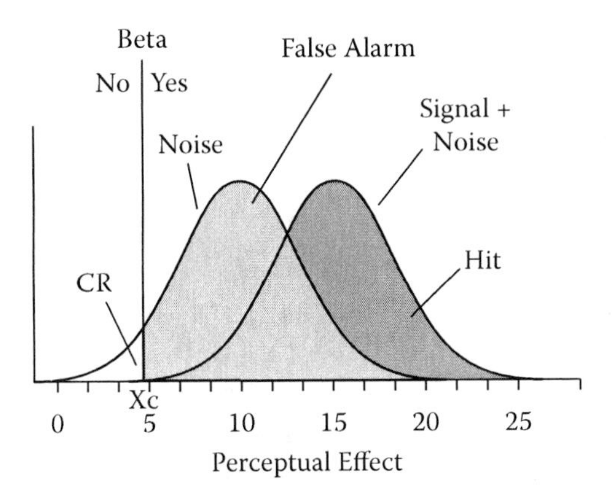

FIGURE 9.3 Liberal bias, greater chance of a positive response.

9.2.2 Discriminability

The amount of noise (internal and external) can be represented as a distribution. The bias also can be represented as a signal distribution. The distance between the means of the noise and signal distributions is the discriminability (d′) measure. This may be estimated through the standard deviations or through the hit and false alarm rate. The greater the d′, the greater the sensitivity in detecting a signal. However, as the level increases, the probability of error also increases. In this regard, SDT has great benefit for quantifying and comparing sensitivity between systems. A balance must be sought between the discriminability of an autonomous system and the sensitivity of the system as measured by d′. Many technologies either in development or presently used to aid manned aircraft pilots are significant steps forward in developing an autonomous DSA system for UASs.

9.3 DETECT, SEE, AND AVOID TECHNOLOGY

9.3.1 Cooperative Technologies

9.3.1.1 Traffic Alert and Collision Avoidance System

The traffic alert and collision avoidance system (TCAS) is the primary cooperative collision avoidance system of a variety of airspace users and transmits information via a transponder. TCAS is considered superior to its predecessor, the traffic advisory system, which provided limited information regarding an intruder aircraft and a possible time of collision (Wolfe 2006). Aircraft designed with a TCAS transponder, permit communication between aircraft in order to avoid collisions. However, one problem with the TCAS system is that aircraft without a transponder may not recognize when other aircraft are near, resulting in a conflict.

Adding weight with a TCAS may be a challenging design issue in UAS. Because of this issue, small UASs may restrict functionality or the inclusion of this system, according to RTCA DO-304. Auditory instructions from TCAS present another problem. Verbal communication between operators, ground pilots, and air-traffic control is already a complex system in UASs and introducing a fourth dimension would complicate things further. In addition, unreliable system automation presents an additional complication (see Doyle and Bruno, 2006).

9.3.1.2 Automatic Dependent Surveillance–Broadcast

Automatic dependent surveillance–broadcast (ADS-B) is a relatively new technology that allows ground-based stations and pilots to detect other similarly equipped aircraft in the airspace. Satellite-based GPS calculates an aircraft's position, altitude, speed, flight number, type, and whether it is turning, climbing, or descending. This information, updated several times per second, then is broadcast to a universal access transceiver, allowing aircraft within roughly a 150-mile radius to see a representation of these signals on a cockpit display of traffic information and ground-based stations to see a representation on regular traffic display screens.

Garmin International Inc. (2007) (a company that has been a leader in global positioning and navigational systems since 1989), Hidley (2006), and Sensis Corp. (2006a, 2006b) outline in detail the many advantages of ADS-B for UAS DSA applications. Briefly, ADS-B provides operators with accurate and reliable information, along with additional navigation variables (e.g., speed, heading, and so forth) at near real time. The extended range of service provided by ADS-B allows more time to avoid collisions. The use of satellites means information is available in locations where radar is ineffective or not available. ADS-B can offer further levels of safety through automated features such as collision warnings. Because it uses proven communication technology, ADS-B can be implemented faster and at a lower cost than other systems. ADS-B also allows for a flexible software structure. This gives it the ability to adapt to future technologies. At the forefront of ADS-B technology is a joint industry and FAA initiative in Alaska known as Capstone. This program has reduced the accident rate in Alaska and demonstrated the effectiveness of ADS-B for UASs.

9.3.1.3 Implications of Using Cooperative Technologies on UASs

Cooperative technologies have a reliable and established track record in terms of reducing midair collisions of manned aircraft; however, they offer several disadvantages for UAS sense and avoid (SAA) systems. SAA systems tend to be cost-prohibitive and only work when all aircraft sharing airspace have an SAA system. They are ineffective against collisions with ground-based obstacles, such as terrain or towers. The SAA systems were developed assuming that a human operator would be involved in each step of system operation (in other words, in the loop), a human would verify warnings, and take appropriate measures. These technologies would most likely require recertification if they were modified for UASs in order to ensure equivalent levels of safety.

9.3.2 Noncooperative Technologies

Some promising technologies being investigated for use in UAS SAA systems are noncooperative technologies, such as radar, laser, motion detection, electro-optical (EO), and infrared (IR). Noncooperative technologies are systems that do not require other aircraft to possess the same technology. Noncooperative technologies benefit from the ability to detect ground-based obstacles as well as airborne.

Noncooperative technologies can be divided into two types: active and passive systems. Active systems, such as radar and laser, transmit a signal to detect obstacles, whereas passive systems, such as motion detection and IR, rely upon the detection of signals emanating from obstacles.

9.3.2.1 Active Systems

9.3.2.1.1 Radar

Radar is an active detection system that uses differences in the time of arrival of reflected electromagnetic waves to create an image of the object(s). Of value to UAS is the synthetic aperture radar (SAR), which uses an integration of multiple radar pulses to create an image. SAR is unique to aircraft because it uses the movement of the aircraft to record data, thereby removing the need for a large antenna. The distance traveled by the vehicle acts as a synthetic aperture and results in a finer resolution than obtained with normal radar (Sandia National Laboratories, 2005). SAR technology is currently being improved to include three-dimensional (3-D) SAR, which uses multiple antennas to create a 3-D image, and change detection, which looks for changes in ground objects by comparing previous images of the same area (Sandia National Laboratories, 2005). Applications of SAR include using SAR for motion detection and to determine location, speed, and size of moving ground targets.

Radar offers several advantages and disadvantages for DSA. Radar systems are ideal in situations where normal optical vision is occluded, such as in inclement weather. Radar pulses can penetrate storms and other weather conditions. However, radar offers disadvantages. These include the large size of traditional radar systems, the high cost, and the fact that radar does not offer the same degree of real-time imagery as an electro-optical system.

Avoiding obstacles in the landscape is the most relevant to a UAS sense-and-avoid application. In this sense, radar can be useful in all phases of flight. Especially considering that the majority of midair collisions occur within 3 miles of the airport, with 50% being below 1000 feet in altitude (Narinder and Wiegmann, 2001).

Sonar uses sound waves in much the same way that radar uses electromagnetic waves. However, due to the slow speed of sound waves relative to the speed of aircraft, the use of sonar technology in UAS sense-and-avoid systems is not ideal (Lee et al., 2004). However, sonar may be useful in small or localized area applications.

9.3.2.1.2 Laser

Laser systems, such as the SELEX Communications Laser Obstacle Avoidance and Monitoring (LOAM®) system, use eye-safe lasers to scan the immediate airspace at regular intervals. The scans are then analyzed using echo-analysis software. Obstacles in the flight path of the aircraft result in a warning to alert operators (SELEX Communications, 2006). Laser systems are currently being used with automated guided vehicles (AGV). These systems scan an area with IR lasers and, upon detecting an obstacle based upon reflected light, can signal the AGV to slow or stop to avoid collision. The systems add an initial cost to the AGV, but the systems reduce maintenance, increase productivity, and lower accident rates (Iversen, 2006).

There are several benefits of a laser-based SAA system, such as the ability to detect nonperpendicular surfaces at high resolution and differentiate between objects as small as 5 millimeters in diameter or as large as buildings (SELEX Communications, 2006). Laser systems are highly configurable, allowing them to compensate for varying atmospheric conditions, thus helping to eliminate the probability of false signal detection.

9.3.2.2 Passive Systems

9.3.2.2.1 Motion Detection

Aircraft can use motion detectors to sense the direction and the velocity of objects. Cameras are placed at different angles to create multiple views, which when combined, can allow for a calculation of object vectors (Shah et al., 2006). These images are compared and if differences in pixel thresholds are met, a vector is calculated for movement. However, during this time, the UAS itself is moving. Numerous companies have developed their own formulas and algorithms to deal with this challenge (Lee et al., 2004; Netter and Franceschini, 2004; Nordberg et al., 2002; Shah et al., 2006). These algorithms cancel out the movement of the UAS as well as noise from the vibration of the UAS itself. The sensors are able to identify objects based on physical characteristics and vectors in order to deal with occlusions.

9.3.2.2.2 Lenslet Model (Insect Model)

An emerging technology utilizes biotechnology with the eyes of flying insects as a model for sensing by attempting to copy the optical flow that is utilized by flying insects (Netter and Franceschini, 2004). Optical flow in insect eyes detects relative motion in contrasts using multiple eye sensors called lenslets. The combined contrast creates patterns that are discerned as movement. Cells in the medulla of the insects, called elementary motion detectors, have been replicated by researchers. These compute the velocity of the contrasts.

9.3.2.2.3 Electro-Optical (EO)

EO systems are sensors that require light for detecting objects. EO systems are limited by the requirement of light as well as being unable to detect intensity or rate of change in intensity of a target. IR type sensors are able to overcome these limitations. Future radar systems that permit the use of radar and EO systems simultaneously are called the active electronically scanned array (AESA) radar-based system (Kopp, 2007). The EO system allows AESA to scan and record imagery while the radar is shifting through its various modes. AESA also has surveillance sensors that can be transformed to reconnaissance sensors by modifying the code. However, the system requires large arrays of antennas and the UAS must carry an additional payload of 3000 pounds. These requirements coupled with a minimum airspeed of 200 knots, effectively makes the UAS a low-flying satellite.

9.3.2.2.4 Infrared

IR technology detects heat in two forms: white-hot objects (WHO) and black-hot objects (BHO). IR requires heat from an object. IR does not require light, which makes it most beneficial for nighttime use. Objects that do not emit heat are colored black or gray in WHO or BHO views, respectively. These onboard sensors have been considered a possible tool for DSA.

9.3.2.2.5 Acoustic

Scientific Applications and Research Associates, Inc. (SARA) developed a compact acoustic sensor system for use on small UASs. The Passive Acoustic Non-Cooperative Collision Alert System (PANCAS) consists of a number of microphones mounted in an array. The PANCAS provides a means of detecting aircraft on a collision course by detecting and tracking the sound of their engines, propellers, or rotors. The PANCAS must deal with random error from atmospheric effects, wind, and signal processing errors. Algorithms are applied to determine a threshold decision for a collision course and to minimize false alarms. Since the system can detect potential collisions from all aspects in azimuth and elevation, collisions can be avoided in situations where a piloted aircraft would normally be blind (e.g., an aircraft being overtaken from behind) (Milkie, 2007).

9.3.2.3 Passive Systems and Ranging

Traditionally, a passive system in a simplistic application may not be thought of as having significant ranging capabilities; however, there have been several concepts developed to deal with this issue. For example, the Air Force Research Laboratory has proposed a solution for detecting range in a passive EO sensor by maneuvering the UAS to establish a baseline with an object and then using the EO sensor to calculate angles and determine range through triangulation. Further research is investigating the type of baseline maneuver that would be required (Grilley, 2005) as well as the ability to eliminate UAS overtaking collisions with noncooperative targets (Kim et al., 2007).

9.3.3 Adapting Technologies

Several technologies that were not developed originally for UAS may be adapted as a potential solution to solve the DSA problem. Using the concept of the compound eye of a fly, researchers at a Swiss laboratory created an optical sensor to avoid fixed obstacle collisions with a very small UAS (Zufferey and Floreano, 2006). This adaptation of a biological sensor is small, simple, and has minimal processing requirements. The 2005 DARPA Challenge Vehicle (Stanford University, 2006), produced by the Stanford Intelligence Laboratory, provides technologies and processing algorithms applicable for UASs flying close to the ground and needing autonomous DSA capability through multiple sensors. Another innovation came from the Brigham Young University-MAGICC UAV (McLain, 2006; Saunders et al., 2005) where an optical computer mouse sensor was adapted to provide a collision avoidance function on a small autonomous UAS. The team at Brigham Young University also used a video camera sensor and range finding laser for a fusion sensor system. These examples demonstrate the wide range of commercially available sensors and technologies applicable to UAS DSA.

9.3.3.1 DSA Demonstration and Testing

The variety of sensors, technologies, and concepts for DSA recently demonstrated or tested showcases the broad assortment of DSA applications developed in academia and industry. The German ATTAS (Friehmelt, 2003) is a full-sized commercial jet that may be used in a "pseudo-UAV" mode to test and analyze systems and procedures. The aircraft can house a wide range of systems to perform test evaluations or exercises for UAS flight applications and DSA applications. The Proteus (see Figure 9.4) is a similarly sized optionally piloted aircraft used in the United States for Skywatch tests (Hottman, 2004; Wolfe, 2002a, 2002b). On the other side of the size spectrum is the 30-gram compound eye Swiss UAV (Zufferey and Floreano, 2006). This tests the ability of a very simple optical sensor to enable obstacle avoidance.

In an effort to test their acoustic-based DSA system, SARA collected the signatures of a number of general aviation piston-engine aircraft and typical turbine-powered helicopters. These signatures were used to test the time it took for detection to occur in a worst-case scenario; two planes traveling toward each other, head on, at maximum speed. Sound delay time and program calculations were collected. The data collected allowed the researchers to calculate the amount of time for which evasive maneuvers were available (Milkie, 2007). In addition to the progress through SARA's research, the Air Force Research Laboratory EO system has gone through a full developmental period, resulting in a sensor system that is suitable for integration into one of the Air Force's high-altitude, long endurance (HALE) UAVs in the near future. Although it does not meet the full requirements for a DSA system for UASs flying in the NAS, it lessens the risk of collision.

Several academic institutions have been involved in the testing of new technologies. Brigham Young University has developed and successfully flown small UASs with innovative, commercially available sensors (Saunders et al., 2005; Theunissen et al., 2005). The Stanford Artificial Intelligence Laboratory's DARPA Challenge

FIGURE 9.4 The Proteus in flight over Rosamond dry lake bed. (Photo by C. Thomas, 2003.)

Vehicle successfully demonstrated the ability to DSA in a ground-based exercise with potential applications to UASs (Stanford University, 2006). Carnegie Mellon University has demonstrated a DSA application in an autonomous helicopter using a TI-C40-based vision system (Carnegie Mellon University, 2010).

Some systems have been developed, tested, and are already in use, such as the Sense-and-Avoid Display System (SAVDS) active ground-based radar system (Zajkowski et al., 2006). These systems increase situation awareness and are suitable for certain small UAS applications to monitor and detect aircraft within a limited range and altitude. This system works for ground-based UAS operators, but is not yet suitable for autonomous UAS.

Systems that have been demonstrated for UAS DSA so far involve a single type of sensor. Several papers (Flint et al., 2004; Suwal et al., 2005; Taylor, 2005) have described concepts of multisensor systems and a mixture of cooperative and noncooperative systems to provide a fuller spectrum of DSA capability. In order for large autonomous UASs to operate in the NAS from takeoff to 60,000 feet to landing, a multisensor system would be required. Performance standards of UAS DSA also need to be developed, as well as full testing facilities, before development of a complete DSA can be accomplished.

9.3.4 Alternative Approaches to Visibility

Other procedural or technological approaches that are not located on the aircraft may enhance the visibility or the awareness of the UAS in the NAS. Currently, much of the visibility research done by the U.S. Department of Defense (DoD) has been focused on military UAS use. The research seeks to reduce the visibility of aircraft to both radar systems and human vision. For UAS operating guidelines and safety with manned aircraft in the NAS, especially civil missions, the desired state of a UAS is high observability. Researchers and developers should note the origin and goals of current visibility research.

9.3.4.1 Electromagnetic Visibility Enhancement

Until 2004, ground-based radar was allowed by the FAA as a means to perform the see-and-avoid function for UAS flights in the NAS. The National Radar Test Facility (NRTF) is tasked with characterizing the radar visibility of aircraft using a variety of electromagnetic test equipment on several size ranges. The DoD's goal, at times, is to minimize or at least understand the aircraft cross-section seen on radar. For civil UAS, the goal would be to increase the cross section for a particular platform. Simplistic approaches would be to minimize the use of radar absorbent material, add reflective edges internal to the platform, and to include designs that are more radar visible when not adversely impacting the aerodynamics.

9.3.4.2 Other Visibility Enhancements

Paint schemes have been used by the DoD to camouflage and decrease the detectability of aircraft, but paint schemes also can be used to increase detectability. Paint schemes can be optimized for different geographies and other environmental factors. In addition, lighting also may be optimized to provide greater awareness to other airspace users of the presence of a UAS. Although the FAA does require UAS to operate with lights on board the platform, the best or brightest lighting with the appropriate transmitting field of view could help the visible detection of the UAS to the human or certain types of DSA sensors.

9.3.4.3 Processes and Procedures for Visibility Enhancement

UAS-specific procedures do exist for specific aircraft operations (e.g., high speed or military training routes) where there would be no opportunity to detect and sense low speed or noncooperative aircraft. These procedures are based primarily on segregation, which provides a mechanical separation of the airspace users. This could mean that UASs are required to operate in prescribed routes for high-density traffic flow between airports. These UAS routes (similar to jet routes) would be a possible procedure for segregating UAS traffic.

Additional research has been done to investigate ways in which to inform air traffic control personnel that a particular aircraft in their sector is a UAS. Hottman and Sortland (2006) modified data blocks and the flight progress strip to add information to the controller so they understood (in the simulation) that something unique existed regarding these particular aircraft (UAS). This methodology is similar to providing information to a controller about a conventional aircraft.

9.4 CONCLUSION

To safely operate UAS in the NAS and minimize the risk of midair collisions, UAS operators must be able to detect and track air traffic to a level of safety equal to or better than that required by the FAA. Most manned general aviation aircraft, which operate under visual flight rules and lack collision avoidance systems, rely on the pilot's eyesight and radio contact with air traffic controllers to track approaching airborne vehicles.

Equipping UAS with TCAS transponders to communicate with other transponder-equipped aircraft reduces the possibility of midair collisions; however, TCAS reduces conflict only with cooperative aircraft. Noncooperative aircraft not equipped with collision avoidance transponders continue to pose a significant risk when flying under visual flight rules. For this reason, UAS operators must be equipped with a sense-and-avoid system that can locate and track both cooperative and noncooperative airborne vehicles at sufficient range to maintain safe separation distances.

Table 9.2 offers a summary of the technological approaches that have been discussed (Hottman et al., 2007). Cooperative technologies for DSA have been developed, tested, and are currently being fielded. The best examples of cooperative technologies are TCAS and ADS-B. These technologies have been developed for manned aircraft with direct applicability for DSA but with capability for UAS.

Some airspace users, such as aircraft, parachutists, and balloons, may not have cooperative systems onboard or systems that are functional. Additionally, ground-based threats to aircraft also exist. In those cases, a noncooperative technology will be required to detect, sense, and avoid these other airspace users although modern GPS dutifully can provide information for terrain avoidance. Cooperative technologies applicable to UAS have an inherent size, weight, and power (SWaP). Depending on the SWaP requirement to operate the cooperative system, this technology may not be applicable to smaller UASs.

There are multiple constraints on noncooperative systems, including environmental, SWaP, and operational constrains. These constraints limit the application of a particular technology based upon the size or class of the UAS. For instance, instrument meteorological conditions (IMC) place an operational limitation on EO technology; whereas day or night places no such restriction. The SWaP on a radar system is generally much higher than technologies such as a camera-based system. Operational limitations also are imposed by the existing design of UAS, which may not accommodate an additional system due to space or configuration. Additional weight could affect center of gravity of aerodynamics and excess power capacity may not be available.

Additional visibility approaches that do not involve adding a technology to the actual UAS platform include paint schemes, lighting, and increasing the radar observability of the platform. Also, segregation of the UAS in the airspace may have significant value. Notifying air traffic controllers through the data block or flight progress strips also would raise the awareness of the type of airspace user.

One single approach may not be adequate for the DSA requirement on a UAS. Where the UAS is operated, its size and the SWaP of the technology along with the technology capabilities need to be considered when determining a DSA suite.

TABLE 9.2
Technology Attributes for DSA

	Detects		Detects		Detects				Constant		Supports		
Technology	Noncoop Targets	Discerns Range	In IMC	Day/ Night	Multiple Targets	Full Time	Multiple Sectors	Detect Range, NM	Azimuth Issue	Co-alt Issue	Due Regard	Asym Covert	Sym Covert
						Search							
Electro-Optical	Yes	No	No	No	??	Yes	Yes	4?	Yes	No	No	Yes	Yes
Human Visual	Yes	No	No	No	No	No	No	2	Yes	No	Yes	Yes	Yes
IR Search & Trk	Yes	Yes	No	Yes	Yes	Yes	Yes	22+	No	Yes	??	Yes	Yes
Passive Infrared	Yes	No	No	Yes	Yes	Yes	Yes	22+	No	Yes	Yes	Yes	Yes
Radar	Yes	Yes	Yes	Yes	Yes	Yes	Yes	22+	No	No	Yes	Yes	No
						Cooperative Surveillance							
TCAS/ACAS	No	Yes	Yes	Yes	Yes	Yes	Yes	22+	No	No	No	Yes	No
ADS-B	No	Yes	Yes	Yes	Yes	Yes	Yes	22+	No	No	No	Yes	No

Source: Hottman, S. B., K. R. Hansen, and M. Berry. 2007. Review of detect, sense and avoid technologies for unmanned aircraft systems (DOT/FAA/AR 08-41). Washington, DC: U.S. Department of Transportation.

Notes: With onboard processing, search technologies may be able to disseminate resolved targets across C2 links. TCAS is a related but separate requirement. ADS-B is in a demonstration phase of development.

ACKNOWLEDGMENTS

Thank you to the Federal Aviation Administration and Stephen B. Hottman for their assistance with this chapter. Portions of the chapter also were printed as Report No. DOT/FAA/AR-08/41.

DISCUSSION QUESTIONS

9.1 Compare and contrast the advantages of light-based detection systems to sound-based detection systems.

9.2 When incorporating a sensor system in a small UAS design, what types of constraints must engineers consider?

9.3 What is the difference between a cooperative sensor technology and a noncooperative sensor technology? What are the advantages and disadvantages of both?

9.4 When selecting or developing a sensor system, what factors must be taken into consideration?

9.5 One novel approach to a multisensor system is the lenslet model based on the biology of an insect eye. What might be another type of multisensory approach based on a biological model?

REFERENCES

Carnegie Mellon University. 2010. The Robotics Institute. http://www.ri.cmu.edu (accessed December 10, 2010).

Doyle, J. M., and M. Bruno. 2006. Predator down: NTSB and U.S. Homeland Security Department look into crash of border-patrolling UAV. *Aviation Week & Space Technology* 164: 35–37.

FAA. 2006. *Aeronautical information manual: Official guide to basic flight information and ATC procedures*. http://www.faa.gov/airports_airtraffic/air_traffic/publications/ATpubs/AIM/.

Flint, M., E. Fernández, and M. Polycarpou. 2004, December. Efficient Bayesian methods for updating and storing uncertain search information for UAVs. 43rd IEEE Conference on Decision and Control, Bahamas.

Friehmelt, H. 2003. Integrated UAV technologies demonstration in controlled airspace using ATTAS. American Institute of Aeronautics and Astronautics. AIAA 2003-5706.

Garmin International Inc. 2007. *ADS-B creates a new standard of aviation safety*. http://www8.garmin.com/aviation/adsb.html (accessed May 21, 2007).

Green, D. M., and J. A. Swets. 1966. *Signal Detection Theory and Psychophysics*. New York: Wiley.

Grilley, D. E. 2005. *Resolution Requirements for Passive Sense & Avoid*. Morgantown, WV: Alion Science and Technology.

Hidley, R.W. 2006, Fall. ADS-B advantages. *Flightline*, no. 3.

Hottman, S. B. 2004. Detect, see and avoid systems for unmanned aerial vehicles. Presentation at New Mexico State University.

Hottman, S. B., K. R. Hansen, and M. Berry. 2007. Review of detect, sense and avoid technologies for unmanned aircraft systems (DOT/FAA/AR 08-41). Washington, DC: U.S. Department of Transportation.

Hottman, S. B., and K. Sortland. 2006. UAV operators and air traffic controllers: Two critical components of an uninhabited system. In *Advances in Human Performance and Cognitive Engineering Research*, ed. N. J. Cooke, 71–88. Bingley, UK: Emerald Group Publishing Limited.

Iversen, W. 2006. Laser scanners for obstacle avoidance. *Automation World*. http://www.automationworld.com/view-1646 (accessed May 21, 2007).

Kim, D. J., K. H. Park, and Z. Bien. 2007. Hierarchical longitudinal controller for rear-end collision avoidance. *IEEE Transactions on Industrial Electronics* 54: 805–817.

Kopp, C. 2007. Active electronically steered arrays. *Air Power Australia*. http://www.ausairpower.net/aesa-intro.html (accessed May 30, 2007).

Lee, D. J., R. W. Beard, P. C. Merrell, and P. Zhan. 2004. See and avoidance behaviors for autonomous navigation. *SPIE Optics East, Robotics Technologies and Architectures, Mobile Robot XVII* 5609-05: 23–34.

McLain, T. W. 2006, December. Autonomy and cooperation for small unmanned aircraft. Proceedings of TAAC Conference in Santa Ana Pueblo, NM.

Milkie, T. 2007. *Passive Acoustic Non-Cooperative Collision Alert System (PANCAS) for UAV Sense and Avoid.* Unpublished white paper by SARA, Inc.

Narinder, T., and D. Wiegmann. 2001. Analysis of mid-air collisions in civil aviation. Proceedings of the 45th annual meeting of the Human Factors and Ergonomics Society.

Netter, T., and N. Franceschini. 2004. Neuromorphic motion detection for robotic flight guidance. *The Neuromorphic Engineer* 1: 8.

Nordberg, K., P. Doherty, G. Farneback, P. Erick-Forssén, G. Granlund, A. Moe, and J. Wiklund. 2002. Vision for a UAV helicopter. Proceedings of IROS'02, Workshop on Aerial Robotics.

Sandia National Laboratories. 2005. *Synthetic aperture radar applications*. http://www.sandia.gov/RADAR/sarapps.html (accessed May 21, 2007).

Saunders, J. B., Call, B., Curtis, A., Beard, R. W., and McLain, T. W. 2005. *Static and dynamic obstacle avoidance in miniature air vehicles*. American Institute of Aeronautics and Astronautics. 2005-6950.

SELEX Communications. 2006. *LOAM®—laser obstacle avoidance system.*

Sensis Corporation 2006a. *Automatic Dependent Surveillance—Broadcast Ground-Based Transceivers.*

Sensis Corporation 2006b. *Capstone: Sensis solutions at work.*

Shah, M., A. Hakeem, and A. Basharat. 2006. Detection and tracking of objects from multiple airborne cameras. *The International Society of Optical Engineering*, 1–3.

Stanford University. 2006. Stanford racing team's entry in the 2005 DARPA Grand Challenge. http://www.stanfordracing.org (accessed May 21, 2007).

Suwal, K. R., W. Z. Chen, and T. Molnar. 2005, September. SeFAR integration TestBed for see and avoid technologies. American Institute of Aeronautics and Astronautics Infotech, Arlington, VA.

Taylor, M. 2005, October. Multi-mode collision avoidance systems (M^2CAS). Proceedings of TAAC Conference, Albuquerque, NM.

Theunissen, E., A. A. H. E. Goossens, O. F. Bleeker, and G. J. M. Koeners. 2005, August. UAV mission management functions to support integration in a strategic and tactical ATC and C^2 environment. AIAA 2005-6310. AIAA Modeling and Simulation Technologies Conference and Exhibit, San Francisco, CA.

Wolfe, R. C. 2002a, October. Cooperative DSA & OTH communication flight test. Proceedings of TAAC Conference, Santa Fe, NM.

Wolfe, R. C. 2002b, October. Non-cooperative DSA flight test. Proceedings of TAAC Conference, Santa Fe, NM.

Wolfe, R. C. 2006. Sense and avoid technology trade-offs. *UAV Systems, the Global Perspective*, 152–155.

Zajkowski, T., S. Dunagan, and J. Eilers. 2006, April 24–28. Small UAS communications mission. Eleventh Biennial USDA Forest Service Remote Sensing Application Conference, Salt Lake City, UT.

Zufferey, J. C., and D. Floreano. 2006. Fly-inspired visual steering of an ultralight indoor aircraft. *IEEE Transactions on Robotics* 22: 137–146.

10 Sensors and Payloads

Douglas M. Marshall

CONTENTS

10.1 INTRODUCTION

It can be safely said that there are two broad categories of remotely piloted aircraft (RPA). The first is the kind that can be seen looping, rolling, and performing entertaining aerobatics, perhaps at a dedicated aero modelers airstrip or out on a deserted field. These types of RPAs are flown recreationally by hobbyists for no other reason than the joy of operating the aircraft and watching it respond to operator commands and doing things that would be impossible or nearly so in a manned aircraft. The second category includes everything else that is flown remotely or without a human pilot on board. These aircraft, rotorcraft, or lighter-than-air blimps are operated for a purpose, and that purpose is to carry an instrument aloft that has some function independent of the navigation or operation of the aircraft. That function is secondary to the physical parameters of flight but primary to the purpose of the flight. In addition to the payload that the aircraft carries, there may be other instruments aboard that generate information for the operator about the health or status of the aircraft, where it is going, how fast and how high, and perhaps assisting the operator in his or her obligation to avoid collisions with other aircraft or persons or property on the ground. This chapter will discuss the history and types of these instruments, their purposes, the regulatory challenges that face the operators, the technological limitations that influence the choice of instruments, and will close by taking a brief look into the future.

10.2 UNMANNED AIRCRAFT SYSTEM (UAS): A "COLLECTION PLATFORM" OR AN AIRCRAFT?

A scientist or researcher desiring to study some phenomenon of nature usually will require the assimilation of some sort of data to analyze. It could be cloud particulate, atmospheric moisture content or pollution, temperatures at the boundary layer, or any number of things that cannot be measured without collecting samples with an instrument or device designed for that purpose. Once the required data set has been identified, a method of collecting the data must be devised, which may require the invention or employment of an apparatus that can meet the scientist's needs. If the most feasible (by technological capability, cost, or resource availability) way to obtain the necessary samples is to get the instrument into the air, and a stationary or ground-based solution is not acceptable, then a choice must be made between unmanned balloons, aircraft or rotorcraft, rockets, or lighter-than-air craft. In the last decade, another alternative has become available to the aviation user community and that is whether the aircraft or flying device needs to have a human on board.

Historically, unmanned aircraft have been used primarily by the military to carry surveillance devices or to deliver weapons. History also suggests that many of the game-changing innovations in aviation have resulted from research and development activities in the defense sector. The successes enjoyed by the military in the use of unmanned or remotely operated systems (ground, air, and water) have informed science, law enforcement, government, academic, and commercial interests looking to remotely piloted systems to perform traditional and newly identified missions in a safer, more efficient, and potentially lower-cost manner.

This evolution of uses for remotely piloted aircraft systems has been stimulated by advances in sensor payload capability and configuration. They can do more with lower power requirements and less weight. Accordingly, the platforms employed to carry these devices have decreased in dimension and weight, and thereby generate a smaller operating footprint. This presents both advantages and disadvantages that will be discussed elsewhere in this chapter.

Users of these systems continually identify new potential applications of the systems and consequently seek enhanced performance capabilities accompanied by lower costs and lesser weight and power requirements. These needs will be described in the following sections.

10.3 REGULATORY CHALLENGES

The status of unmanned systems with respect to their ability to operate in the National Airspace System (NAS) remains unsettled pending comprehensive overhaul or amendment by the Federal Aviation Administration (FAA) (see Chapter 3). Perhaps the greatest challenge facing UAS developers and users are the requirements of the Federal Aviation Regulations (FARs) right-of-way rules. The right-of-way rule states that "when weather conditions permit, regardless of whether an operation is conducted under instrument flight rules or visual flight rules, vigilance shall be maintained by each person operating an aircraft so as to see and avoid other aircraft.

When a rule of this section gives another aircraft the right-of-way, the pilot shall give way to that aircraft and may not pass over, under, or ahead of it unless well clear."* The "see and avoid" phrase creates the technical challenge of providing an onboard sensor that offers the same capability for the UAS operator to observe conflicting air traffic or ground obstacles as the human in an occupied aircraft. Much work and research is being conducted to solve that particular problem with a technical solution.

The airspace operating rules of Part 91 require by inference that any civil aircraft operating in the NAS be properly registered and flown by a certificated pilot ("flight crew member").† Part 61 states that a person may not act as a pilot in command or any other capacity as a required pilot or flight crew member of a civil aircraft of U.S. registry unless that person meets certain regulatory requirements.‡ Aircraft operating in the NAS under the authorization of a U.S. airworthiness certificate must also be maintained in compliance with specific requirements (excepting experimental aircraft).§ Many other sections of the FARs may have specific application to UAS, but that is a discussion for another chapter.¶ Since there is no pilot on board a UAS, the sensor payload may perform double duty by providing imaging for remote sensing and perhaps being available for some level of sense-and-avoid capability, thereby providing an alternate means of compliance with the relevant sections of parts 61 and 91. The FAA has not certified any onboard see-and-avoid or sense-and-avoid system that would relieve the operator of the obligation to either provide a chase plane or ground observers to keep the UAS in the line-of-sight for conflict avoidance purposes. Such a device, if and when developed and certified, would be a navigation and pilotage tool for the UAS, and most likely not serve simultaneously as a payload that is intended to collect data. However, due to size, weight, and power constraints, efforts may be underway to combine both functions into one device. It is important to distinguish between the two technologies and their purposes when considering payloads and sensors.

10.4 SENSORS AND PAYLOADS: IS THERE A DIFFERENCE?

A discussed earlier, a sensor on a UAS may perform any number of functions that are intended to facilitate the mission of the aircraft (the "collection platform"). Recreational UASs (model aircraft) typically do not have any sensor on board, and under current FAA guidelines they may not be so equipped if the sensor (usually a camera) is used to collect images or other data for commercial purposes.** Neither *sensor* nor *payload* is defined in the FARs.†† For purposes of understanding the unique characteristics of unmanned or remotely piloted systems, and the missions they are intended to execute, it is instructive to consider a payload as a descriptor of a suite or configuration of sensors on the one hand, or the capability of the aircraft

---

* 14 CFR Part 91.113(b).
† 14 CFR Part 91.1(a).
‡ 14 CFR Part 61.3.
§ 14 CFR Part 43.1.
¶ See Chapter 3.
**FAA Interim Operational Approval Guidance 08-01.
†† 14 CFR Part 1.1.

to carry and deliver a dispensable load such as fire retardant or crop spraying chemicals. In the military arena, of course, payload on a UAS or RPA usually means some type of armament. Payloads may be viewed as the reason for the flight of the aircraft, and the absence of a payload does not render the aircraft incapable of flight. In that context, sensors can be seen to collect data and remain with the aircraft, whereas payloads may refer to product that leaves the aircraft and is delivered or dropped. However, the two terms have commonly used interchangeably to describe sensors as devices that are designed to meet the demands of the mission.

The operator or proponent of a certificate of authorization or waiver (COA) typically will first identify the scope or goal of the mission. What data is to be collected? Where is it best collected, in what atmospheric conditions, and under what environmental constraints? What is the optimum device to do that job? UAS designers will endeavor to create collection platforms that can fill as many roles as possible, which may involve integrating several sensors into one sensor package that can readily be "plugged" into a payload bay. A "chicken or the egg" situation develops, where the issue becomes whether the sensor payload should be developed or identified first and then either fit into an existing platform, or should the platform be designed around the sensor package and the mission? Another approach is to start with an available platform that possesses the endurance, altitude, and mass carrying capacity that are required, and then collect the sensors that meet the mission needs and also fit into the platform. Cost constraints will usually drive the user toward compromises between all those factors to best achieve the mission goals while keeping the project within affordable limits.

The sensor payload types fall into five broad categories: (1) navigation and safe transit; (2) communication and control; (3) remote sensing and imaging, (4) intelligence, surveillance and reconnaissance; and (5) data collection (air samples, moisture and pollution, particulates, temperature).

10.5 SENSE-AND-AVOID DYNAMICS AND SYSTEMS

As discussed earlier, operations in the NAS require compliance with the relevant sections of Part 91 of the FARs, particularly Sections 91.111 Operating Near Other Aircraft* and 91.113 Right-of-Way Rules: Except Water Operations.† In the visual flight rules (VFR) environment the pilot has primary responsibility for safe separation from other aircraft. In instrument conditions, air traffic control provides separation services, but when the weather allows (clear of clouds or in daylight hours), the pilot still has the duty of vigilance for other aircraft. The terms *see and avoid* and *well clear* are not defined in the FARs, although the Aeronautical Information Manual (AIM) describes a *near midair collision* as "an incident associated with the operation

* 14 CFR 91.111(a): "No person may operate an aircraft so close to another aircraft as to create a collision hazard."

† 14 CFR 91.113(b): "General. When weather conditions permit, regardless of whether an operation is conducted under instrument flight rules or visual flight rules, vigilance shall be maintained by each person operating an aircraft so as to see and avoid other aircraft. When a rule of this section gives another aircraft the right-of-way, the pilot shall give way to that aircraft and may not pass over, under, or ahead of it unless well clear."

of an aircraft in which the possibility of collision occurs as a result of proximity of less than 500 feet to another aircraft, or a report is received from a pilot or a flight crew member stating that a collision hazard existed between two or more aircraft."*

The pilot or operator of any aircraft flying in the NAS is obligated to avoid operating his her aircraft so close to another aircraft as to "create a collision hazard," a phrase that has not been clearly defined, but, at the very least, means not to come within 500 feet, laterally or vertically, of another aircraft.† A *near miss* is a reportable incident (reportable to the FAA).‡ In addition to the duty to avoid flying too close to another aircraft, pilots are also obligated to maintain vigilance so as to see and avoid other aircraft. *See and avoid* is another term not well-defined nor well-understood but can reasonably be interpreted to mean that pilots are obligated to maintain "situational awareness" and be constantly on the lookout for other aircraft, even if operating under instrument flight rules where air traffic control is providing traffic advisories and separation. The Aeronautical Information Manual states that the pilot's responsibility "when meteorological conditions permit, regardless of the type of flight plan or whether or not under control of a radar facility" is to be responsible for seeing and avoiding other traffic, terrain, or obstacles.§ Of course, common sense dictates that pilots would want to do this anyhow, purely as a matter of survival, as few midair collisions are ever nonlethal.

Undoubtedly one of the most challenging barriers to the ability of unmanned or remotely piloted aircraft to fly in unrestricted (off military restricted or warning areas) airspace is the see-and-avoid requirement. Although the question of the capability of humans to accurately and safely see and identify a moving target in the air through the windscreen of an airplane is far from settled, a remotely piloted aircraft certainly does not have even that capability without some sort of sensor or visualization system on board that can serve as an acceptable substitute. The FAA's stated goal or standard for a certified sense-and-avoid system is a "target level of safety" equivalent to 1×10^{-9}, or one fatality per 1 billion flight hours (system-wide), which is the standard for commercial aviation.¶

The industry response to these requirements has been to explore a wide variety of solutions to essentially provide the unmanned system operator or pilot with a suite of sensors that can provide a visual capability equivalent to or better than the human eye. More important than an "electronic eyeball" is the need for the cognition of a human who is not only able to see the target but to process the information received to make rapid and real-time determinations about the relative bearing, azimuth, altitude, speed, and physical characteristics of the target so as to make a decision about the necessity of an evasive maneuver.

Ongoing research efforts and flight experiment results have been focused on over-the-hill surveillance and urban reconnaissance operations, where the images or packets of data generated by the onboard sensors provide intelligence to the operators as

* Aeronautical Information Manual § 7-6-3.

† FAA Airman's Information Manual (AIM) § 7-6-3 Near Midair Collision Reporting.

‡ AIM, supra.

§ AIM § 5-5-8.

¶ 14 CFR 23.1039(a) and (b) and FAA Advisory Circular 23.1309-1D.

they navigate in the battlefield environment. The question is: Does this technology have potential to address the see-and-avoid requirement as well?

The challenge to developers and researchers is to create a system or system of systems that has the ability to ensure that the remotely piloted aircraft can operate safely near other aircraft when outside of controlled airspace. Once the RPA sensors have detected another aircraft as a potential collision threat, the RPA pilot typically has little time to respond. Before certification for an airborne see-and-avoid system will be granted, the FAA will require a fail-safe solution to guarantee avoidance. One suggested strategy is to combine the features of radar and an electro-optical sensor to generate visual images to provide the equivalent of human sight. This could include a pilot-in-the-loop as the cognitive maneuver decision maker using an aircraft collision avoidance system (ACAS) algorithm.

Going beyond the foregoing scenario, researchers want to determine how to make such a system more autonomous, allowing the unmanned aircraft to generate more aggressive maneuvers to avoid close-in contacts without exceeding structural and aerodynamic limits. The perfect solution would address all the myriad operational, policy, regulatory, public perception and technical issues that now prevent routine integration of RPAs into civil airspace. The problem is that no one sensor provides an RPA operator with all the information needed to avoid a collision. Video data provides angular information, both in altitude and in azimuth, so the airplane or target object (perhaps a bird) can be seen, but gives no distance information or the closure rate between the RPA and the target.

Other sensors, like laser range finders, offer more precise distance information, but they do not give good angular information (height and direction) unless it is a fully scanned laser range finder. And, laser range finders are too heavy for most RPAs (the most common small aircraft or rotorcraft contemplated for use in the NAS weigh less than 55 pounds and have a payload capacity of less than 15 pounds).

To be eligible for certification by the FAA as a sense and avoid, Part 91-compliant collision avoidance system, the device should be evaluated by way of simulation across millions of randomly generated close encounters that represent actual operations in the NAS. New encounter models that capture changes that have occurred in U.S. airspace since earlier models were developed in the 1980s and 1990s are continually being revised and updated. These models accurately simulate the characteristics of small, general aviation aircraft that may not be utilizing air traffic control services as well as typically larger aircraft that are transmitting a discrete transponder code. Newer encounter models allow for dynamic (physics-based) changes in airspeed, vertical rates of climb, and turn rates that were not previously possible.

Thus, to meet the FAA's system safety requirements, UAS developers have proposed a number of different onboard sensors. The list may include the traffic alert and collision avoidance system (TCAS), which is mandated for all commercial aircraft authorized to carry more than 19 passengers, automatic dependent surveillance–broadcast (ADS-B), infrared (IR) and electro-optical (EO) systems, radar, laser, and acoustic systems. TCAS and ADS-B provide a satisfactory means of sensing cooperative (transponder-equipped) aircraft but lack the ability to detect targets that are not equipped with the proper avionics ("uncooperative targets" such as parachutists, sail planes, hot air balloons, and flocks of birds). EO, IR, and radar sensors are attractive solutions for detecting traffic because they do not require that other aircraft or potential targets have special equipage. EO and IR systems have the advantage of power requirements and payload sizes that are smaller than radar systems.

Other sense-and-avoid strategies contemplated by developers include optic flow sensors, laser range finders, acoustic sensors and onboard computer or synthetic vision devices that have the potential to produce "equivalent level of safety" capabilities as integrated systems (where the data from each device are fused via complex algorithms into an image or display equivalent to what a human could see and react to). Ultimately, these integrated system designs may form the foundation for autonomous systems that require no human intervention and offer even greater see-and-avoid capabilities than human cognition.

More detailed descriptions of various sense-and-avoid systems are provided in Chapter 9 and will not be repeated here. The point of this discussion is to compare and differentiate navigation-oriented technologies from purpose-driven sensors that are either peripheral to sense-and-avoid systems or entirely independent from them except to the extent that they may share power sources and compete for communications bandwidth.

10.6 PURPOSE-DRIVEN SENSORS

As distinguished from navigation-related devices, purpose-driven sensors are those that are intended to support the mission of the collection platform. The list of potential uses for unmanned systems seems to grow by the day or week as government agencies, research and academic institutions, scientists, law enforcement, and others seek substitutes for or supplements to the existing tools in their toolboxes. Any activity or data set acquisition that is currently being achieved by manned aviation, satellites, balloons, or even other methods that do not involve aerial technology is a potential candidate for UAS application. In the military and law enforcement environments, the most popular uses are daylight and low-light level cameras that produce both still and video images for intelligence and surveillance. Thermal imaging cameras operate in the infrared (heat radiation) wavelength spectra, permitting interpretation of images at night and low visibility conditions. Image quality is degraded by extended rainfall because rain lowers the temperature of inanimate objects, thereby reducing contrast with heat-producing targets such as humans, animals, or internal combustion engines.

In the civil aviation arena, remote sensing cameras, whether infrared, electro-optical, or multispectral, offer images of agricultural fields, forests, and grasslands that permit analysis of moisture content, pestilence, and other influences on the health of the observed flora. EO and IR cameras mounted on gimbaled and stabilized platforms provide marine or land mammal researchers nonintrusive images of their subjects, which are difficult to obtain with manned aircraft or ground observers. Small UASs can operate at relatively low altitudes for long periods of time while producing high-resolution imaging with a very small noise footprint that minimizes chances of scaring animals into the water or out of visual range. Other proposed uses (yet to be approved by the FAA due to the Part 91.113(b) see-and-avoid issues discussed earlier) include pipeline and power line inspection using video and multispectral cameras, surveillance and interdiction for law enforcement, search and rescue operations arising out of natural disasters, severe weather events, lost and missing persons, and other situations where high-resolution cameras on stabilized platforms with long endurance capabilities can supplement or substitute for manned aerial assets. Small or miniaturized synthetic aperture radar packages have been employed on UASs for ground surveillance. In these cases, the sensors and cameras are truly payloads, carried aloft by collection platforms that are designed especially for the purposes stated.

The scientific community has enthusiastically adopted UAS platforms and a wide variety of sensors specially designed to gather data in an ever-growing list of environments and applications. Climate researchers employ remote sensing and data collection technologies on both small (less than 50 pounds) and large (greater than 15,000 pounds) UASs to collect data on chemical plumes, air pollution, volcanic ash, temperatures at the boundary layers, wind and moisture content near hurricanes and tornados, ice melting rates and composition, overland river flooding, and pollution monitoring. UASs are also useful for satellite calibration and validation, since they can stay aloft for many hours in preprogrammed flight plans that reduce the opportunities for human error due to fatigue and boredom that are inherent in manned aviation operations. The list of uses for UASs to carry sensor packages and payloads continues to expand and is only limited by the imagination and innovative thinking of scientists, researchers, and UAS developers, and of course the willingness of the regulators to permit such operations in the NAS.

10.7 TECHNOLOGICAL AND SYSTEM LIMITATIONS

The ability of unmanned aerial systems to fulfill their missions depends in large part upon the communications link between the UAS and the ground control station (GCS). The communications link is a two-way highway, consisting of an uplink that allows the operator or pilot stationed at the GCS to send commands to the aircraft, and the downlink that sends data back to the GCS. The returned data may include the images or other information collected by the payload, the health status of the payload and the aircraft itself, and possibly a communications link to another receiving station. The uplinks and downlinks through the UAS may also serve as communications links between the operator and air traffic control, particularly when the UAS is operated beyond line-of-sight (BLOS). Currently, the only method of linking UASs to the GCS is radio communication, either by direct line-of-sight transmission (C-band radios in the 4 to 8 GHz range) or, in BLOS operations, via satellite (K_u-band in the 12 to 18 GHz range). Bandwidth is at a premium worldwide, and efforts are under way to secure frequencies for UAS operations.* Increasing demand for access to the radio spectrum for commercial (primarily television and radio), scientific research, and other interested parties has made frequency allocation for UAS a top priority in the research and development organizations that are making substantial investments in UAS technology.

* World Radiocommunication Conference, organized every 4 years by the International Telecommunication Union: http://www.itu.int/ITU-R/index.asp?category=information&rlink=rhome&lang=en.

Competing needs arising from economic, technological, scientific, and industrial interests will challenge the World Radiocommunication Conference delegates and frequency managers to find room in the limited available spectra for UAS operations. Conflicting frequency allocations between regions and countries compel UAS developers and exporters to be especially diligent in researching frequency availability in locations outside of the United States. For line-of-sight operations, the limitations are as technical as they are regulatory. Frequency managers must secure permissions (licensing in most cases) for discrete frequencies to use for UAS operations, and the range will vary from 60 to 100 miles, depending upon the operating altitude of the UAS and the antenna height of the radio transmitter. Currently, the FAA does not grant COAs for operations beyond a 5-mile visual range, although proponents routinely request authority for flights well beyond that limit.

All payload configurations bear the same size, weight, and power burden, even for the larger military or science-based UAS (see Chapters 9 and 12 for more detailed discussion.). Some sensors require a great deal of power, and designers must consider the implications of the sensors running off the same power source as the other electrically powered instruments or the motor itself if the UAS is electric motor-driven. Others, such as infrared cameras or lasers, may require cooling, which adds to the weight burden as well as power requirements. Many of the more popular UAS, both in military and civilian use, are relatively small (less than 55 pounds) and thus cannot carry payloads much heavier than 10 to 15 pounds. The goal is typically to carry the most capable payloads with the longest endurance to satisfy the demands of the mission, so subsystem integration and miniaturization become more important, often leading to single-sensor modular payload designs that can be readily changed between missions. Other operators may require multiuse sensors that can perform a variety of functions that do not necessitate swapping out payload packages as the mission changes (see Chapter 12 on miniaturization).

Of course the issue that is equal to the technological challenges is certification and standardization. Currently, no payload configuration or modular assembly has been certified as acceptable or essential equipment for UASs operating in the NAS. There are no standards that specifically address UAS sensors or payloads. Although the biggest efforts have been directed toward detect, sense, and avoid technology to enable "file and fly" access to the NAS, sensor packages are routinely used by the science community to achieve some of the research goals described earlier. The aircraft employed in these operations are flown under the auspices of an FAA issued COA or in protected, military, or government controlled airspace. These operations stimulate further design refinements and generate data for the FAA's demand for comprehensive safety cases that will justify further expansion of existing UAS activities.

10.8 CONCLUSION

The terms *sensor* and *payload* are commonly used interchangeably, but in reality they may mean slightly different things. A suite of sensors (electro-optical, infrared, multispectral cameras) may be integrated into one payload. Many commercial and military UASs have payload bays that allow for interchangeable packages of sensors or cameras. Other, perhaps larger UASs, may have separate bays, sensor balls, and wiring for onboard radar or other optical devices that provide forward-looking views for piloting and navigation or for surveillance. As UASs become more capable, with longer endurance and higher altitude limits, operators will continue to seek more extensive access to the NAS, particularly beyond line-of-sight operations with autonomous systems that require little or no operator intervention once the aircraft is launched. Large sensor or optical devices that provide the data collection capability or finer resolution sought by researchers and eventually commercial operators are being miniaturized and designed for to draw less power so that endurance and payload capacity are not degraded. Other payloads may be dispensable, such as dispersants or pesticides, and present entirely different challenges in UAS design and concepts of operation. Strategically, the UAS mission designer or planner will want to first determine what information is to be gathered, then identify the sensors or suite of sensors that best accomplishes that goal, then seek out the platform that is best suited to carry out the mission, given size, weight, power and endurance limitations. Ultimately, the choice of sensor payloads and platforms will be greatly influenced by cost and the ability of the integrated system to be operated safely in the National Airspace System under the oversight of the regulators, as compared to manned assets that are capable of executing the same tasks.

DISCUSSION QUESTIONS

10.1 Discuss the two categories of RPA.
10.2 Briefly describe the differences between sensors and payload.
10.3 Discuss the see-and-avoid challenges facing UAS operations.
10.4 Define purpose-driven sensors.

11 Human Factors in Unmanned Aircraft Systems

Igor Dolgov and Stephen B. Hottman

CONTENTS

11.1 INTRODUCTION

The discipline of human factors emerged primarily from aviation requirements during World War II. Pioneers of human factors science, such as Alphonse Chapanis and Paul Fitts, extended the principles of aviation psychology to human–machine interaction. This changed the research and design approach from one that treated humans and machines as separate entities to one that treated it as a human–machine system. The change afforded researchers the possibility to empirically evaluate various facets of human–machine systems and make appropriate recommendations to interface designers. Today, the areas of human factors and ergonomics (the nomenclature can be regional) include the disciplines of psychology, anthropometry, ethnography, engineering, computer science, industrial design, operations research, and industrial engineering. Moreover, in the later part of the 20th century, human factors engineering principles have been applied to product and industrial design on a worldwide scale, in both the industrial and government sectors.

In the past two decades, human factors science has come full circle and is widely applied in the context of unmanned aircraft systems (UASs). The prior 10 chapters of this volume introduce the reader to the history, regulations, procedures, design and engineering of UAS. The variability present in UAS platforms is evident in the content of the foregoing chapters. Hundreds of historical and modern platforms originating from more than 40 countries across the globe may be found with an online search. Numerous systems are employed by the U.S. Armed Forces, with different branches instituting proprietary multitiered classification schemes, loosely based on operational altitude (Air Force), operational distance (Army), and aircraft size (Navy and Marines).

The variability in UASs allows them to operate within many military, commercial, and civil applications. Many UAS operations teams consist of a mission planner, internal pilot, external pilot, and payload operator. In an automated, large-scale UAS such as the Air Force's Global Hawk, the mission planning plays a pivotal role in comparison to real-time control. On the other hand, small and micro UASs, like the Bat (Air Force), the Wasp (Navy and Marines), or the Raven (Army), require the presence of an external pilot for some flight phases during takeoff and landing. Some UASs also may require an internal pilot to complete the remainder of the mission remotely.

Another reason for the observed variability in UASs is that operational context and task-relevant goals motivate the overriding pragmatic concerns, which are the primary driving force behind the design of UASs. The resulting systems inevitably exhibit differences in operator demands and workload, entailing variable human factors across existing platforms, as evident in the following analyses of military UAS accidents and mishaps.

11.2 UNMANNED AIRCRAFT SYSTEM ACCIDENT AND MISHAP ANALYSIS

In 1996, the Air Force Scientific Advisory Board (AFSAB) identified the human–machine interface as the facet of UASs that needed the most improvement (Worch et al., 1996). The most reliable evidence to the causes of UAS mishaps comes from the U.S. military. Human error has accounted for roughly half of all UAS mishaps. This ranges between 28% to 79% across U.S. military forces (see Table 11.1), and 21% to 68% across UAS type (Rash et al., 2006). In 2001, the Department of Defense (DoD) estimated that UASs suffered accidents at a rate of 10 to 100 times that of what is observed in manned aircraft, with operator error accounting for approximately 20% of all mishaps (Department of Defense, 2001).

An insightful comparison of accident data across the branches was made by Rash et al. (2006). As shown in Table 11.1, there is some variability in human involvement in UAS accidents among the branches of the U.S. military. Rash et al. argue that the operational context and level of UAS automation at takeoff and landing is connected to the disparity in human error rates in the Air Force compared to the other branches. The UAS accident data gathered from the Air Force is gathered from combat operations employing a highly automated UAS (e.g., the Predator). The Army and Navy data are gathered from training sessions with UASs that employ an external pilot. Rash et al. conclude that the inflated human error rates observed in the Air Force are a result of mismatched automation (Parasuraman

TABLE 11.1
Summary of UAS Human Error Rates in UAS Accidents and Mishaps across U.S. Military Branches

Source	Time Period of UAS Operations	Number of Accidents Analyzed	U.S. Air Force (%)	U.S. Army (%)	U.S. Navy/ Marines (%)
Manning et al. (2004)	1995–2003	56	—	32	—
Schmidt and Parker (1995)	1986–1993	107	—	—	33
Rogers et al. (2004)	1993–2003	48	69	—	—
Seagle (1997)	1986–1997	203	—	—	31
Williams (2004)	1980–2004	12, 74, 239	67	36	28
Tvaryanas et al. (2005)	1994–2003	221	79	39	62

Source: Rash, C., P. LeDuc, and S. Manning, 2006, Human errors in U.S. military unmanned aerial vehicle accidents. In *Human Factors of Remotely Operated Vehicles,* ed. N. Cooke, H. Pringle, H, Pedersen, and O. Connor, 117–132, Oxford, UK: Elsevier.

and Riley, 1997) and the added contribution of stress experienced in field operations compared to training. Error rates also are higher for manned aircraft operations for takeoff and landing phases.

However, both perspectives fail to adequately account for the apparent differences between the branches. In reality, the Air Force's fully automated Predator system presents less opportunity for human intervention than the Army and Navy systems that utilized an external pilot, a scenario that Parush (2006) has been shown to have the greatest propensity for accidents, particularly during training. Moreover, although operations in the context of training are less stressful, UAS trainees are novices and should be expected to make more errors in training than the experts deployed in the field due to their lack of experience.

The U.S. Air Force Research Lab Report #2004-11 by Manning et al. (2004) was the most detailed of the analyses listed in Table 11.1. The report compared the human factors analytics and classification system (HFACS) and 4Ws inventory (what, when, why, what; see later for further explanation) approaches to classifying accidents. The HFACS classifies four levels of failure: level 1, preconditions for unsafe acts; level 2, unsafe acts; level 3, unsafe supervision; and level 4, organizational influences. These global categories are further decomposed into 17 causal factors of operational error, as shown in Figure 11.1 (Shappell and Wiegmann, 2000; Wiegmann and Shappell, 2003).

Although there are numerous error taxonomies (Adams, 1976; Bird, 1974; Degani and Weiner, 1994; Firenze, 1971; Helmreich and Foushee, 1993; O'Hare et al., 1994; Sanders and Shaw, 1988; Suchman, 1961), Reason's (1990) systems approach to nuclear accident analysis has revolutionized the way mishaps and accidents are modeled. Reason's model is referred to as a Swiss cheese model because of latent and active failures occurring at different levels in the hierarchy of the system in which the error occurred (see Figure 11.2). The HFACS account of such mishaps addresses

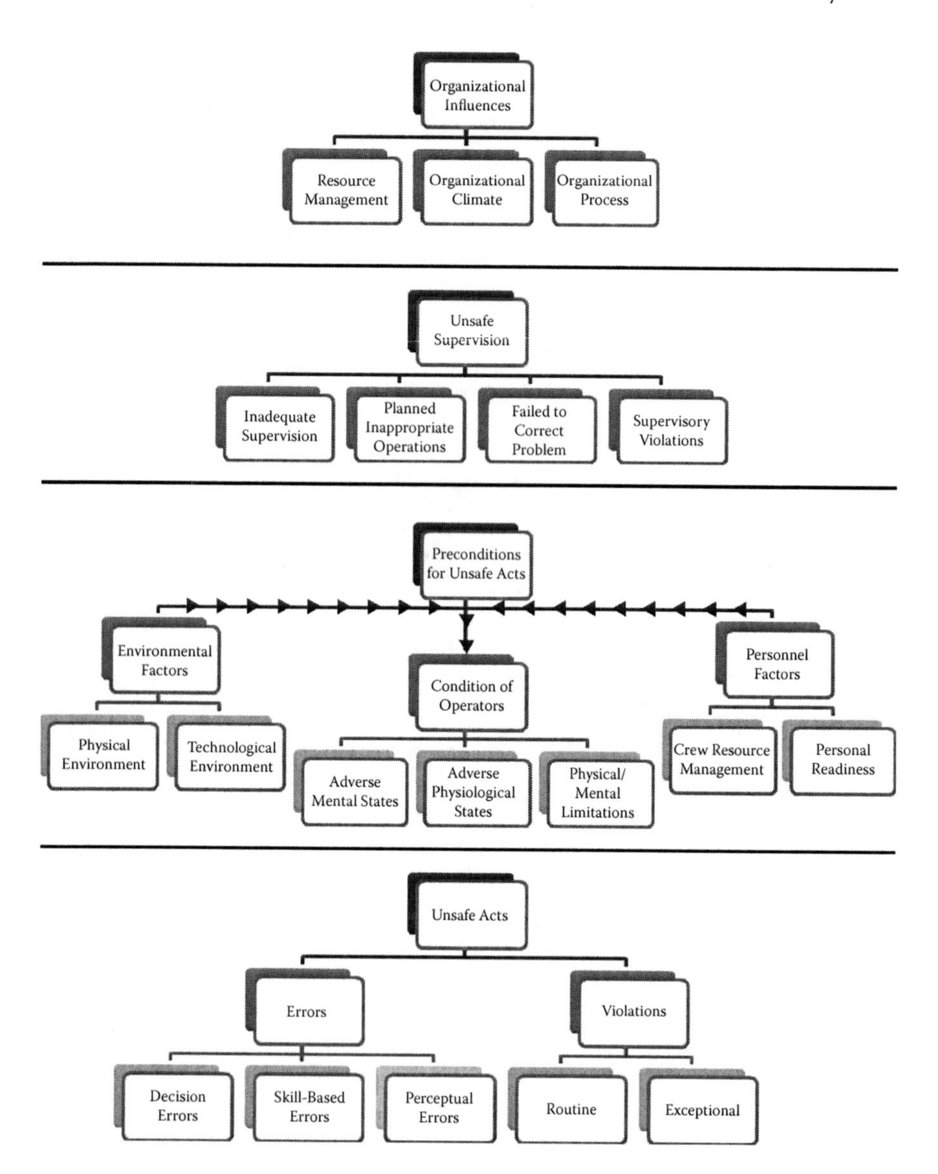

FIGURE 11.1 The four global categories (organizational influences, unsafe supervision, preconditions for unsafe acts, and unsafe acts) and 17 causal human factors of error specified by the HFACs. (Adapted from Wiegmann, D., and A. Shappell, 2003, *A Human Error Approach to Aviation Accident Analysis: The Human Factors Analysis and Classification System*, Hampshire, UK: Ashgate.)

a few shortfalls of Reason's model and builds an architecture of mishap analysis that has been validated numerous times.

The HFACS is a multifaceted inventory and draws directly upon cognitive, behavioral, ergonomic, organizational, industrial, and aeromedical perspectives. Since its inception in the 1990s, it has been employed successfully by Wiegmann

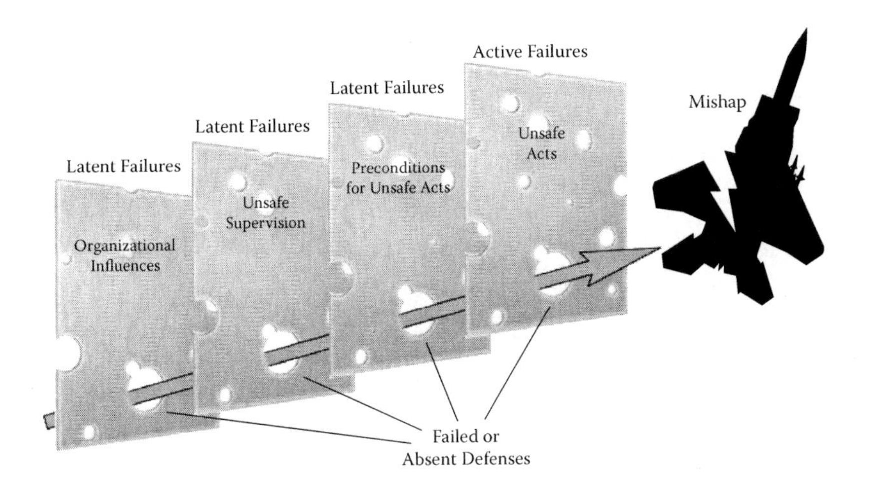

FIGURE 11.2 Reason's Swiss cheese model of safety failures. Errors trickle down as a result of latent failures at higher levels of the hierarchy all the way down to individual unsafe acts. (Adapted from Wiegmann, D., and A. Shappell, 2003, *A Human Error Approach to Aviation Accident Analysis: The Human Factors Analysis and Classification System*, Hampshire, UK: Ashgate.)

and Shappell (2003) in a number of military and civilian contexts. In the context of UASs, it has also been successfully employed by Manning et al. (2004); Tvaryanis and Thompson (2008); and Tvaryanis et al. (2005).

Manning et al. (2004) found that the UAS errors attributed to humans fell into the following areas: unsafe acts (61%), unsafe supervision (50%), and organizational influences (44%). The data were then analyzed using the 4Ws approach and compared to the results of the HFACS analysis. The 4Ws approach is based on the methodology laid out in Department of the Army Pamphlet 385-40 and addresses the sequence of events that led to the eventual failure by asking the following four questions: (1) When did the error/failure/injury occur? (2) What happened? (3) Why did it happen? (4) What should be done about it? After these questions are answered, failures are categorized as follows: individual failure, leader failure, training failure, support failure, and standards failure (Department of the Army, 1994).

The results of applying the HFACS and 4Ws were congruent and identified human error as the mishap cause in the same 18 (of a total of 56) instances. Furthermore, the authors found correlations between the four levels of failure specified in the HFACs and the five categories of failure of the 4Ws approach suggesting the relationships depicted in Figure 11.3. Manning et al. (2004) lamented the lack of attention paid to adequate reporting procedure and the unavailability of demographic data for a significant proportion of the sample.

Tvaryanis et al. (2005) improved on the methodology used by Manning et al. (2004) and conducted a 10-year cross-sectional analysis of human factors in U.S. military UAS mishaps using the HFACS system. It was determined that 60% (133

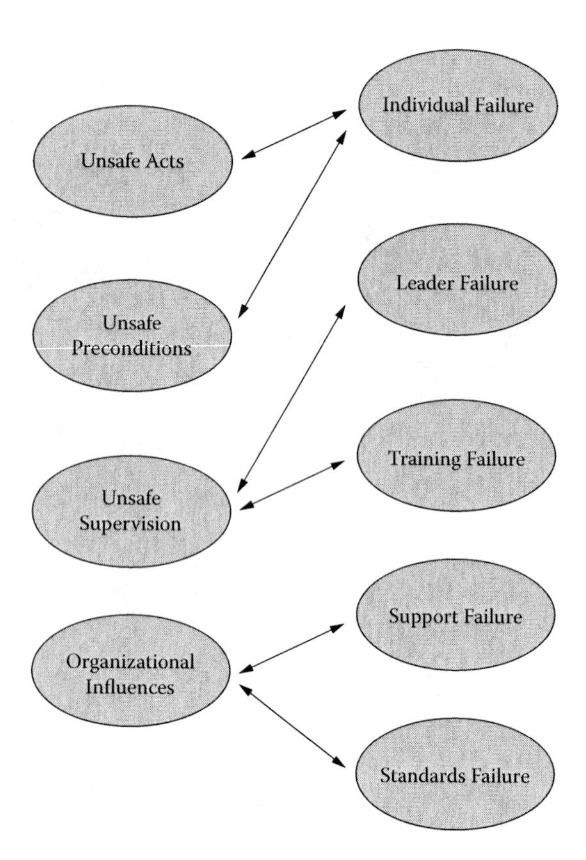

FIGURE 11.3 Suggested relationships between error causal factors of the HFACS and Army Pamphlet 385-40 found in the Manning et al. (2004) study. (Adapted from Rash, C., P. LeDuc, and S. Manning, 2006, Human Errors in U.S. Military Unmanned Aerial Vehicle Accidents, In *Human Factors of Remotely Operated Vehicles,* ed. N. Cooke, H. Pringle, H, Pedersen, and O. Connor, 117–132, Oxford, UK: Elsevier.)

of 221) of the analyzed incidents were a result of human intervention. Models predicting unsafe operator acts were constructed using binary logistic regression. Technological environment and cognitive factors ($p < 0.010$) were found to be predictors of unsafe acts in the Air Force; in the Army, organizational processes, psychobehavioral factors, and crew resource management ($p < 0.001$) were predictors; and in the Navy, work and attention and risk management ($p < 0.025$) predicted unsafe acts. Skill-based errors were the most common in the Air Force ($p = 0.001$), and violations were the most common in the Army ($p = 0.016$) when compared to the other services. The authors concluded that recurring latent failures at the organizational, supervisory, and preconditions levels contributed to more than 50% of UAS mishaps across the branches, but the pattern of errors differed between the branches of the Armed Forces. The authors suggested different interventions for each branch based on these findings.

Tvaryanis et al. (2005, 2008) and the studies reported in Tables 11.1 and 11.2 are evidence of the great impact that operational context has on UAS performance.

Moreover, failures occur at every level of the UAS operation, from the operator in the field to the highest organizational levels. Such wide differences suggest the consideration of systematic examination of each UAS set to take flight in U.S. airspace. The Government Accountability Office (GAO) recommends that such case-by-case analysis should occur at research facilities designated by the Federal Aviation Administration as capable of in-depth human factors and systems-level evaluation, such as the New Mexico State University UAS Flight Test Center (GAO, 2005).

11.3 HUMAN FACTORS OF UNMANNED AIRCRAFT SYSTEMS

During the first UAS Technical Analysis and Applications Center (TAAC) conference in 1999, a variety of issues related to UASs, including human factors, were addressed by an international audience. The first academic workshop on UASs focused on cognitive issues was sponsored by the Cognitive Engineering Research Institute (CERI) at Arizona State University. Along with the selection of formal talks and posters, the conference participants organized into a number of breakout groups centered on the following themes: cognition and perception, selection and training, simulation displays and design, team process, and systems safety (Connor et al., 2006). McCarley and Wickens (2004) proposed a similar taxonomy: displays and controls; automation and system failures; and crew composition, selection, and training.

The aforementioned divisions along with the categories utilized by the taxonomy of unsafe acts and the HFACS (Wiegmann and Shappell, 2003) comprise the set of human factors and related human–system components that are most important for UAS operations. Analyses of military and civilian UAS accident reports indicate that performance cannot be reduced to a simple subset of factors. Moreover, the impact of human factors on system performance is greatly mitigated by operational context, human system integration, system automation, crew composition, and crew training.

11.3.1 Operational Context

The exact demands on the human–machine interface are primarily defined by the operational context. As detailed in Chapter 3 through Chapter 6, UAS operations must abide by regulations that provide guidance within the mission context. A wide range of UASs have been designed to fill varied niches in military, law enforcement, civilian, and academic domains. Operational context directly impacts operator workload and situational awareness, and has implications for crew size and the desired degree of system automation (Cooke et al., 2006). Although it is a general aim of all interface designers to minimize or distribute workload, such goals are particularly desirable in life-critical domains, where operators are pressured to make decisions and execute procedures effectively and efficiently.

TABLE 11.2
Summary of Prior UAS Mishap Studies

Source	U.S. Military Branch	Taxonomy Used	Percent of Accidents That Involved Human Factors	Human Factors and Corresponding Percentage of Their Involvement in Mishaps
Schmidt and Parker (1995)	Navy, $n = 170$	None	>50% (estimated)	Aeromedical screening
				Selection procedures
				Crew resource management (CRM)
				Crew station design
				Career field development
Seagle (1997)	Navy, $n = 203$	Taxonomy of unsafe acts	43%	Unsafe acts (59%)
				Accidental acts (52%)
				Slips (2%)
				Lapses (16%)
				Mistakes (39%)
				Conscious acts (7%)
				Infractions (7%)
				Unsafe condition (46%)
				Aeromedical (20%)
				CRM (27%)
				Readiness violations (7%)
				Unsafe supervision (61%)
				Unforeseen (34%)
				Foreseen (47%)
Ferguson (1999)	Navy, $n = 93$	Taxonomy of unsafe acts	59%	Unsafe acts (38%)
				Intended (17%)
				Mistakes (12%)
				Violations (7%)
				Unintended (20%)
				Slip (14%)
				Lapse (3%)
				Unsafe condition (40%)
				Aeromedical (10%)
				CRM (28%)
				Readiness violations (10%)
				Unsafe supervision (43%)
				Unforeseen (15%)
				Foreseen (12%)
Manning et al. (2004)	Army, $n = 56$	HFACS	32%	Unsafe acts (61%)
				Skill-based (22%)
				Decision (33%)
				Misperception (17%)
				Violations (11%)

(continued)

TABLE 11.2 (continued)
Summary of Prior UAS Mishap Studies

Source	U.S. Military Branch	Taxonomy Used	Percent of Accidents That Involved Human Factors	Human Factors and Corresponding Percentage of Their Involvement in Mishaps
				Preconditions (6%)
				CRM (6%)
				Unsafe supervision (50%)
				Inadequate supervision (33%)
				Failed to correct known problem (17%)
				Supervisory violations (11%)
				Organizational influences (44%)
				Organizational processes (44%)
Rogers et al. (2004)	Air Force, Army, $n = 48$	Human–systems issues	69%	Training (27%)
				Team performance (25%)
				Situational awareness (18%)
				Interface design (16%)
				Cognitive and decision making (14%)

Source: Tvaryanas, A., W. Thompson, and S. Constable, 2005, *U.S. Military Unmanned Aerial Vehicle Mishaps: Assessment of the Role of Human Factors Using Human Factors Analysis and Classification System (HFACS)* (Technical Report HSW-PE-BR-TR-2005-0001), Washington, DC: General Printing Office.

11.3.2 Human–Systems Integration

The *Defense Acquisition Guidebook*, published online in 2004, stresses a systems approach in various U.S. federal projects. Chapter 6 in the guidebook is dedicated to human–systems integration (HSI) and recommends applying a user-centered approach to systems design and deployment throughout.

HSI is a systems-level analysis of complex organized behavior that involves the practical integration of engineering, human factors and ergonomics, personnel, manpower, training, crew composition, environment habitability and survivability, and system safety perspectives (DoD, 2003, 2004). "The fundamental concept underlying HSI is the consideration of the human element in all aspects of a system's life-cycle so as to reduce resource utilization and system costs from inefficiency while dramatically increasing system performance and productivity" (Tvaryanis et al., 2008, p. 2).

Human factors such as perceptual abilities, cognitive capacity, situational awareness, and the ability to perform under stress or in high cognitive-demand situations contribute to the effectiveness of the human–machine system. McCauley (2004) identifies six HSI issues that will continue to apply to the domain of UAS: (1) human roles, responsibilities, and level of automation; (2) command and control/concept of operations; (3) manning, selection, training, and fatigue; (4) difficult operational environments; (5) procedures and job performance aids; and (6) moving control platforms. Efficiency and efficacy cannot only be maximized by improving the user but also through improving the link between the human and the system. Displays, controls, and the overall human–machine interface design are the component of HSI that compliments the user. By applying user-centered design techniques and ensuring that ergonomic principles are instituted in real-world systems, designers can create UASs that can fulfill their true potential (DoD, 2005). However, as McCauley (2004) observes, although ergonomic design principles are the short term-goal, the introduction of highly successful automation should be the long-term goal of UAS interface designers.

11.3.3 System Automation

As evident in Chapter 7 in this volume, UASs differ widely on the degree of automation that is built into the system. Systems with a high-degree of automation, like the Global Hawk, also carry extensive mission planning overhead requiring every detail of the mission to be planned in advanced, a process that can take several days. On the other hand, man-portable UASs are typically deployable in minutes.

The obvious benefit of automating operational procedures is to reduce operator workload, thereby improving situational awareness and allowing crew members to attend to mission-critical tasks as they arise. However, UAS autonomy is often imperfect. The unreliability leads to a loss of trust by interface operators, especially in instances where the operator's assessment of the situation is not in line with the automation (Parasuraman and Riley, 1997). Furthermore, people differ in their rates of agreement with an automated aid when it states that a critical event has occurred (compliance) than when it states the opposite (reliance) (Schwark et al., 2010). Although the use of automation in UAS has been increasing, it has yielded only small benefits in performance. Future work in this field demands an empirical assessment of the usefulness of automated aids.

11.3.4 Crew Size, Composition, and Training

Crew size and composition can have a significant impact on the ability of the operations teams to execute their mission. Although it is possible to operate some small and micro UASs alone, most systems require a flight crew consisting of several members (external and/or internal pilot), a payload/reconnaissance operator, and sometimes a mission planner (Cooke et al., 2006). The ability of the crew to communicate and pass the control of the vehicle from one member to another has an enormous impact on UAS performance (McCarley and Wickens, 2004; Williams, 2004). To combat mission failure in such situations, operation teams should train for less-than-optimal situations as well as mundane scenarios.

In this modern age of weapon systems that have varying levels of autonomy and other attributes, are basic piloting skills the most important considerations for training effectiveness, efficiency, and mission success? According to Deptula (2008) "The most important tactical skill Airman [*sic*] will need in the 21st Century will be the ability to rapidly acquire, develop, and share information across the Joint Force, and at all levels of warfare." When weapons are involved, there must be a higher level of judgment and accountability (Deptula, 2008).

New UASs and technologies do not fit neatly into the currently accepted training programs, and these flight systems are being produced faster than the existing flight-training regime can react to them. Furthermore, new technology systems that perform similar functions do not look alike and operator interactions with these systems often are completely different (Hottman and Sortland, 2006).

A consensus or phenomenological (versus empirical) based approach has been in place for four separate technical standards–related aviation bodies that have addressed UAS operator requirements. The ASTM F38 recommendation is based on the "Type" certification model (Goldfinger, 2008), parallel to manned aircraft, where the pilot receives a license based on a specific unmanned aircraft and a specified pilot position within that UAS platform. Some positions would require FAA commercial or instrument ratings and some would not. The FAA in 2010 entered into a relationship with ASTM to formulate a variety of UAS-related standards. The SAE worked for over 4 years on a candidate training syllabus and has developed a primary syllabus and a subset of the primary syllabus based upon specific parameters of the UAS mission (Adams, 2008). Additionally, RTCA Inc. has active endeavors with the FAA to address a variety of UAS technologies and subsystems including the operator. Last, the FAA small UAS (sUAS) Aviation Rulemaking Committee (ARC) is actively seeking the establishment of rules for sUASs.

NATO also has worked on UAS operator standards resulting in a recently approved standardized agreement (STANAG) (2006). For the designated UAS operator, the STANAG lists skills by subject knowledge areas, task knowledge, and task performance (knowledge, skills, and abilities [KSAs]). These KSAs are tailored to UAS type and role (NATO, 2006).

The FAA's (2008) qualification requirements for UAS pilots primarily depend on the flight profile, size, and complexity of the UAS and whether the flight operation occurs near a public airport. Although each COA can establish specific limits, the FAA nearly always requires the pilot in command of the UAS to possess a private pilot's license for any operation conducted above 400 feet aboveground level (AGL) or at distances greater than 1 mile from the pilot. Since the great majority of UAS are designed to operate beyond line of sight, operators essentially must provide a certificated pilot for all but the most basic UAS operations. The FAA also requires UAS operators to provide a certificated pilot for operations in nearly all conditions in which they might encounter a manned aircraft. The primary users of UAS to date are the armed forces although consistent practices have not been evident across the services. The introduction of UASs to more civil and commercial applications can benefit from the military experience but a number of empirical issues remain when considering the civil regulatory agency responsibilities (Hottman and Hansen, 2007; Hottman and Zaklan, 2007).

11.4 CONCLUSION AND FUTURE RESEARCH DIRECTIONS

Human factors analysis and user-centered design are a critical part of every contemporary human–machine enterprise. In the domain of UASs, more than a decade of research has resulted in a large collection of potential human factors that could influence UAS performance. Although UAS accident rates can partially be attributed to imperfect aircraft propulsion and control designs, the reliability of UASs needs to improve about two orders of magnitude to compare to manned aircraft (Office of the Secretary of Defense, 2003; Office of the Under Secretary of Defense for Acquisition, Technology, and Logistics, 2004). As the HFAC's analyses of Manning et al. (2004) and Tvaryanas et al. (2005, 2008) demonstrate, human factors errors occur as a result of latent failures at the organizational and supervisory levels. Organizational and supervisory failures may lead to unsafe preconditions that produce active failures and unsafe acts. By addressing human factors at each of these levels, UAS safety and performance can be significantly improved.

Human factors and training are closely associated for a number of human–machine systems and UASs are no different in this association. Depending on the level of autonomy, the operator moves toward more of a system monitor than as a direct human-in-the-loop operator. Vigilance can and has become an issue in UAS operations; how well the human trusts the automation, workload, situational awareness, and other human–systems issues all have an impact on training. The ongoing debate about whether the overall KSA for operating a UAS should require pilot certification will continue along with balancing the skill set needed for the operationally relevant mission, such as remote sensing.

A recent report estimates that the U.S. military UAS market is projected to grow at a compound annual growth rate of 10% from 2010 to 2015, equaling approximately $62 billion (Market Research Media, 2010). Yet, as the most recent GAO report states the DoD is quickly racking up an impressive UAS fleet without robust plans for personnel, facilities, and the communications structure to support them (GAO, 2010). The GAO advised federal agencies (U.S. military branches, Department of Homeland Security, and FAA) to cooperate to ensure UAS safety and expand its potential uses within the NAS (GAO, 2005). The same report's recommendations for executive action to the FAA included two missions: (1) finalizing the issuing of a UAS program plan to address the future of UAS, and (2) analyzing the data the FAA collects on UAS operations under it COAs and establishing a process to analyze DoD data on its UAS research, development, and operations. In spite of similar objectives having been set out in the DoD UAS roadmaps (DoD, 2001, 2005), these goals have only been partially met to date.

Although the majority of recent UAS development has been in the military sector, UASs in nonmilitary contexts are becoming popular. Eleven commercially available UASs are now being sold in the United States (McCarthy, 2010; Wise, 2010). Although nonmilitary operations of UASs present challenges similar to military systems, there are additional regulatory concerns as noted by the European Joint Aviation Authorities UAV Task Force Report (2004). The FAA is developing an integrated roadmap for the introduction of military, public service, commercial and civilian UAS into the U.S. national airspace. These guidelines are expected to be in place by early 2020. Research in the human factors of UAS flight will continue to play a critical role in ensuring the safe proliferation of such systems through the United States and the world. With careful data gathering and recording techniques (Manning et al., 2004) and the use of widely applicable metrics, such as the HFACs (Wiegmann and Shappell, 2003), UAS researchers and designers will ensure that these systems achieve the level of performance that is needed for this technology to reach its full potential.

DISCUSSION QUESTIONS

11.1 In which contexts, phases of operation, and for which types of UASs do you think human factors play the greatest part in ensuring operational efficiency and efficacy?

11.2 How do you think the needs of civilian and commercial UASs differ from those of the military?

11.3 What kind of challenges do you anticipate in operating UASs at night, dusk, or dawn as opposed to daytime operations?

11.4 What would be the most important differences between operating UASs in rural versus urban areas? How could UASs be tailored to meet the challenges of both environments?

11.5 Name a few previously untapped application areas in which you could see benefit from the implementation of UASs. Consider the costs and benefits of employing UASs in these contexts.

11.6. A number of nations have developed their own UAS platforms. What do you foresee as the biggest challenges to the global integration of UASs?

REFERENCES

Adams, E. (1976). Accident causation and the management system. *Professional Safety* 21: 26–29.

Adams, R. (2008, December). *UAS Pilot Training Recommendations*. UAS TAAC 2008 Conference, Albuquerque, NM.

Bird, F. (1974). *Management Guide to Loss Control.* Atlanta, GA: Institute Press.

Cooke, N., H. Pedersen, O. Connor, and H. Pringle. (2006). CERI human factors of UAVs: 2004 and 2005 workshop overviews. In *Human Factors of Remotely Operated Vehicles*, ed. N. Cooke, H. Pringle, H. Pedersen, and O. Connor, 3–20. Oxford, UK: Elsevier.

Cooke, N., H. Pringle, H. Pedersen, and O. Connor (Eds.). (2006). *Human Factors of Remotely Operated Vehicles.* Oxford, UK: Elsevier.

Degani, A., and E. Weiner. (1994). Philosophy, procedures, and practice: The four P's of flight deck operations. In *Aviation Psychology in Practice*, ed. N. Johnston, N. McDonald, and R. Fuller, 44–67. Brookfield, VT: Ashgate.

Department of the Army. (1994). *Army Accident Investigation and Reporting* (Army pamphlet 385-40). Washington, DC: General Printing Office.

Department of Defense. (2001). *Unmanned Aerial Vehicles Roadmap, 2005–2025.* Office of the Secretary of Defense. Washington, DC: General Printing Office.

Department of Defense. (2003). *Department of Defense Instruction 5000.2: Operation of the Defense Acquisition System.* http://www.dtic.mil/whs/directives/corres/pdf/500002p.pdf (accessed May 18, 2010).

Department of Defense. (2004). *Defense Acquisition Guide (2004).* https://dag.dau.mil/Pages/Default.aspx (accessed May 18, 2010).

Department of Defense. (2005). *Unmanned Aircraft Systems (UAS) Roadmap, 2005–2030.* Office of the Secretary of Defense. Washington, DC: General Printing Office.

Deptula, D. (2008, December). The indivisibility of intelligence, surveillance, & reconnaissance (ISR). UAS TAAC 2008 Conference, Albuquerque, NM.

FAA. (2008). Unmanned aircraft systems operations in the U.S. National Airspace System. Interim Operational Approval Guidance 08-01.

Ferguson, M. (1999). Stochastic Modeling of Naval Unmanned Aerial Vehicle Mishaps: Assessment of Potential Intervention Strategies. Master's thesis, Naval Postgraduate School, Monterey, CA.

Firenze, R. (1971). Hazard control. *National Safety News* 104: 39–42.

Goldfinger, J. (2008, December). ASTM International Committee F38 on unmanned aircraft systems. UAS TAAC 2008 Conference, Albuquerque, NM.

Government Accountability Office. (2005). Unmanned aircraft systems: Federal Actions needed to ensure safety and expand their potential uses with the national airspace system (Technical report GAO-08-511). http://www.gao.gov/products/GAO-05-511 (accessed May 23, 2010).

Government Accountability Office. (2010). Unmanned aircraft systems: Comprehensive planned and a results-oriented training strategy are needed to support growing inventories (Technical report GAO-10-331). http://www.gao.gov/products/GAO-10-331 (accessed May 23, 2010).

Helmreich, R., and H. Foushee. (1993). Why crew resource management? Empirical and theoretical bases of human factors training in Aviation. In *Cockpit Resource Management*, ed. E. Weiner, B. Kansi, and R Helmreich, 3–45. San Diego, CA: Academic Press.

Hottman, S. B., and K. R. Hansen. (2007, June). UAS operator requirements research. Presented at the UVS UAV Conference #9, Paris, France.

Hottman, S., and K. Sortland. (2006). UAS operators and air traffic controllers: Two critical components of an uninhabited system. In *Human Factors of Remotely Piloted Vehicles*, ed. N. Cooke, H. Pringle, H. Pedersen, and O. Connor. Oxford, UK: Elsevier.

Hottman, S., and D. Zaklan. (2007, October). UAS research and civil airspace applications. Briefing for UAVNET Alliance Meeting #16, Madrid, Spain.

Manning, S., C. Rash, P. LeDuc, R. Noback, and J. McKeon. (2004). The Role of Human Causal Factors in US Army Unmanned Aerial Vehicle Accidents (Technical report USAARL Report No. 2004-11). Wright-Patterson AFB, OH: U.S. Air Force Research Laboratory.

Market Research Media. (2010). *U.S. Military Unmanned Aerial Vehicles (UAV) Market Forecast 2010–2015.* http://www.marketresearchmedia.com/2010/04/09/unmanned-aerial-vehicles-uav-market/ (accessed May 18, 2010).

McCarley, J., and C. Wickens. (2004). Human factors concerns in UAV flight. http://www.hf.faa.gov/docs/508/docs/uavFY04Planrpt.pdf (accessed January 13, 2010).

McCarley, J., and C. Wickens. (2005). Human factors implications of UAVs in the national airspace (Technical report AHFD-05-05/FAA-05-1). University of Illinois, Aviation Human Factors Division.

McCarthey, E. (2010). Civilian UAVs: Five more aircraft. *Popular Mechanics*. http://www.popularmechanics.com/science/space/4213471 (accessed May 23, 2010).

McCauley, M. (2004). *Human Systems Integration and Automation Issues in Small Unmanned Aerial Vehicles.* Thesis, Naval Postgraduate School, Monterey, CA.

NATO. (2006). *STANAG 4670. Recommended Guidance for the Training of Designated Unmanned Aerial Vehicle Operator (DUO).* Brussels: NATO/NSA.

Office of the Secretary of Defense. (2003). *Unmanned Aerial Vehicle Reliability Study*. Washington, DC: Department of Defense. http://www.acq.osd.mil/uav/ (accessed January 13, 2010).

Office of the Under Secretary of Defense for Acquisition, Technology, and Logistics. (2004) *Defense Science Board Study on Unmanned Aerial Vehicles and Uninhabited Combat Aerial Vehicles*. Washington, DC: Department of Defense. http://www.acq.osd.mil/dsb/reports/uav.pdf (January 3, 2005).

O'Hare, D., M. Wiggings, R. Batt, and D. Morrison. (1994). Cognitive failure analysis for aircraft accident investigation. *Ergonomics* 37: 1855–1869.

Parasuraman, R., and V. Riley. (1997). Humans and automation: Use, misuse, disuse, abuse. *Human Factors* 39: 230–253.

Parush, A. (2006). Human errors in UAV takeoff and landing: Theoretical account and practical implications. In *Human Factors of Remotely Operated Vehicles,* ed. N. Cooke, H. Pringle, H, Pedersen, and O. Connor, 91–104. Oxford, UK: Elsevier.

Rash, C., P. LeDuc, and S. Manning. (2006). Human errors in U.S. military unmanned aerial vehicle accidents. In *Human Factors of Remotely Operated Vehicles,* ed. N. Cooke, H. Pringle, H, Pedersen, and O. Connor, 117–132. Oxford, UK: Elsevier.

Reason, J. (1990). *Human Error.* New York: Cambridge University Press.

Rogers, B., B. Palmer, J. Chitwood, and G. Hover. (2004). Human–systems issues in UAV design and operation (Technical Report HSIAC-RA-2004-001). Wright-Patterson AFB, OH: Human Systems Information Analysis Center.

Sanders, M., and B. Shaw. (1988). *Research to Determine the Contribution of System Factors in the Occurrence of Underground Injury Accidents.* Pittsburg, PA: Bureau of Miners.

Schmidt, J., and R. Parker (1995, July). Development of a UAV mishap human factors database. Proceedings of Association for Unmanned Vehicle Systems International Unmanned Systems Conference, Washington, DC.

Schwark, J., I. Dolgov, W. Graves, and D. Hor, (2010, September). The influence of perceived task difficulty and importance on automation use. Human Factors and Ergonomics Society Conference, San Francisco, CA.

Seagle, J. (1997). Unmanned aerial vehicle mishaps: A human factors analysis. Thesis, Embry-Riddle Aeronautical University Extended Campus, Norfolk, VA.

Shappell, S., and D. Wiegmann. (2000). The Human Factors Analysis and Classification System—HFACS (Technical report DOT/FAA/AM-00/7). Office of Aviation Medicine, Federal Aviation Administration, Department of Transportation.

Suchman, E. (1961). *A Conceptual Analysis of Accident Phenomenon, Behavioral Approaches to Accident Research.* New York: Association for the Aid of Crippled Children.

Tvaryanas, A., and W. Thompson. (2008). Recurrent error pathways in HFACS: Analysis of 95 mishaps with remotely piloted aircraft. *Aviation, Space, and Environmental Medicine 79*: 525–531.

Tvaryanas, A., W. Thompson, and S. Constable. (2005). U.S. *Military Unmanned Aerial Vehicle Mishaps: Assessment of the Role of Human Factors Using Human Factors Analysis and Classification System (HFACS)* (Technical report HSW-PE-BR-TR-2005-0001). Washington, DC: General Printing Office.

Tvaryanas, A., W. Thompson, and S. Constable. (2008). The U.S. military unmanned aerial vehicle (UAV) experience: Evidence-based human systems integration lessons learned. In *Strategies to Maintain Combat Readiness during Extended Deployments: A Human Systems Approach, 5-1–5-24,* RTO-MP-HFM-124, Paper 5. Neuilly-sur-Seine, France: RTO.

Wiegmann, D., and A. Shappell. (2003). *A Human Error Approach to Aviation Accident Analysis: The Human Factors Analysis and Classification System*. Hampshire, UK: Ashgate.

Williams, K. (2004). *A Summary of Unmanned Aircraft Accident/Incident Data: Human Factors Implications* (Technical report DOT-FAA-AN-04-24). Oklahoma City, OK: Civil Aerospace Medical Institute, Federal Aviation Administration.

Wise, J. (2010). Civilian UAVs: No pilot, no problem. *Popular Mechanics*. http://www.popularmechanics.com/science/space/42134641 (accessed May 23, 2010).

Worch, P., J. Borky, R. Gabriel, W. Hesider, T. Swalm, and T. Wong. (1996). *U.S. Air Force Scientific Advisory Board Report on UAV Technologies and Combat Operations* (Technical report SAB-TR-96-01). Washington, DC: General Printing Office.

12 The Future of Unmanned Aircraft Systems

Richard K. Barnhart

CONTENTS

12.1 INTRODUCTION

Heraclitus of Ephesus (c. 535 BC–475 BC), an ancient Greek philosopher, was noted for his observance of the constancy of change in the universe. Nowhere is that constancy more evident than in the high-tech world of unmanned aerial systems (UASs). In fact, it is likely that in the fairly near future, the term *UAS* will become obsolete in favor of a more publicly acceptable term such as *remotely piloted aircraft* (it is becoming evident that the term *unmanned* is becoming a liability when it comes to public policy formulation related to something that flies) (Deptula 2010). As such, writing about the future of UASs is a bit slippery so the majority of this chapter will focus on a 3- to 5-year timeline where the concepts are more certain.

12.2 ANTICIPATED MARKET GROWTH

No discussion of an industry is complete without an examination of the historical and future trends of that industry. It has only been within the last 10 to 15 years or so that UASs, formerly called UAVs (unmanned aerial vehicles) or drones, have been referred to as an industry—that makes it a young industry by any measure. As such, it has grown steadily from being a barely noticeable segment of the aerospace industry to being a major segment in a relatively short span of time driven by technological enablement. As is often the case, new segments of industry experience rapid growth, almost immune from the economic cycles that affect more mature industries. This has certainly been, and continues to be, the case with the UAS market.

When we refer to UAS market expenditures, the market can be broken down into eight basic segments:

- Research, development, testing, and evaluation (RDT&E)
- Platforms or air vehicles
- Ground control systems
- Payloads and sensors
- Service and support
- Sensor data processing and dissemination
- Training and education
- Procurement, public and private

By all accounts the worldwide UAS market is forecast to experience strong growth. Two major UAS market research firms both predict tremendous market growth in the UAS segment over the next 5 to 10 years. Teal Group Corporation, a team of integrated market research analysts, estimates in its 2010 forecast that UAS expenditures worldwide will top $80 billion over the next 10 years with the United States accounting for 76% of the research and development portion of those funds and 58% of the procurement dollars (Figure 12.1) (Aboulafia, 2010). It also reports that the UAS sector of aerospace manufacturing continues to be the most dominant sector in terms of growth. Expenditures are expected to grow from $4.9 billion annually to $11.5 billion annually by the year 2020. Expenditures are simply dollars that are directed to all eight segments of the UAS market.

MarketResearch.com is another group that predicts very strong growth in the UAS market. Its numbers are a bit more optimistic in that it predicts that the UAS market will reach $63 billion by the year 2015 ("U.S. Military Unmanned Aerial Vehicles," n.d.). MarketResearch.com describes this growth as meteoric.

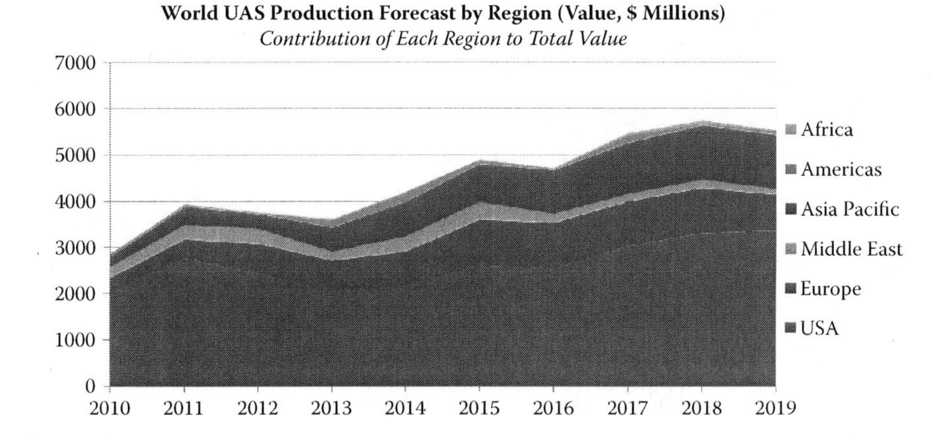

FIGURE 12.1 Teal Group, World UAV systems 2010 market profile and forecast. (Reprinted with permission.)

12.2.1 Private

The private sector market for UAS is the dam that is ready to break. Currently, there are significant restrictions placed by the Federal Aviation Administration (FAA) on private companies seeking to operate UASs in the National Airspace System (NAS). The current mechanism for most UASs to operate in the NAS is to apply to the FAA for a certificate of authorization (COA), which establishes a wide spectrum of controls and limitations for specified UAS operations. Current procedures call for a public entity to sponsor any COA that is issued in the United States, which private companies see as an undue burden. Market pressure is resulting in changes to these restrictions and it seems that in the near future (within 3 years) there will be a regulatory structure for private UAS operation in the NAS. Once in place, this will allow for private contractors to offer services for a wide range of UAS services to include surveillance (which will bring its own set of challenges), air monitoring, communications relays, and airborne messaging.

12.2.2 Public

This segment of the market will continue to be dominated by the military, law enforcement, and university-related research activity. The military continues to lead the way into the future by continuing to push the limits of technology to meet the needs of commander situational awareness and they are responsible for the vast majority of current UAS sector spending. Other public sector initiatives will continue to grow as well as law enforcement and first responders take advantage of regulatory enablement and technological advances, and universities advance the limits of technology through research.

12.3 INFRASTRUCTURE

12.3.1 Ground-Based

Infrastructure can be defined as the physical structures and the service or organizational framework that allows an industry, organization, or society to function in an organized way. As an example, when we refer to our national transportation infrastructure we are referring to our roads, bridges, rail lines, water and airways, harbors, and airports on the physical side, and on the services side we refer to the associated servicing and maintenance organizations to include the associated training and education.

Questions have been raised that with so much anticipated growth in the UAS industry, is the current aviation infrastructure capable of absorbing this onslaught of air vehicles? Certainly we now have challenges in terms of available infrastructure but the good news is that there is much opportunity and room for future growth. As an example, in the United States we enjoy an expansive network of underutilized general aviation airports that are well connected to the NAS, many of which are located out of the main air traffic flow. From this runway access, the UAS infrastructure has room to expand to meet future market growth.

As an example, the Las Cruces (New Mexico) International Airport has taken advantage of ample infrastructure and compatible surrounding airspace to embrace the UAS industry through its relationship with the Physical Science Laboratory's Technical Analysis and Application Center (TAAC) associated with New Mexico State University. At this facility UAS platforms are tested and evaluated for flight readiness before being released for routine operations.

Another example of this is in Herington, Kansas, which has created the Herington UAS Flight Facility (HUFF) at the local general aviation airport to do just that. Its first COA, issued in the spring of 2010, was for the CQ-10A powered parachute UAS. The HUFF, located on a former military airfield from WWII, is an example of a general aviation airport, which, like many others across our nation, has experienced a decline in traffic volume in recent years as the industry struggles with the economic downturn. This has created opportunities for new industries such as UASs. The HUFF's mission is to enable UAS testing, evaluation, and operation through its relationship with Kansas State University's Unmanned Aerial Systems Program Office.

Ironically, an important piece of the current ground-based infrastructure necessary for full UAS integration will be the implementation of the FAA's NextGen NAS modernization plan, which shifts the NAS system from being ground-based (permanent navigation aid equipment, radar, VHF communications, etc.) to satellite-based. In this system, aircraft position and other relevant information are widely available, thereby enabling all users to make better informed air traffic decisions. Ultimately the hope is for UAS aircraft to be enabled to make independent collision threat assessments and take evasive action, which is coordinated with surrounding aircraft through a satellite-based system.

12.3.2 Routine Airspace Access

As the capability of UAS platforms increase and they become more affordable, there is growing demand to allow these vehicles to fly routinely in the NAS outside the special issuance process of obtaining a COA. The COA process has existed as a temporary measure to enable UAS operations in the NAS while the FAA determines exactly how to best integrate UASs into the NAS in response to growing demand from operators and potential customers. As of now it seems as if routine UAS operations in the NAS will become a reality within a few short years where UAS will be able to file an instrument flight rules (IFR) flight plan and operate in a manner similar to their manned counterparts. Exactly how these operations will be allowed will be specified in the forthcoming UAS Federal Aviation Regulations (FARs), which is now in the rulemaking process. This rule will specify a structure for allowable UAS operations in the NAS by operational weight and performance. The rule will be similar to other FARs in that it will specify operational limitations and requirements for the vehicles and operators.

12.3.3 Training and Certification

For many years, federal standards have governed all facets of aviation from operator training and certification requirements to materials certification, and manufacturing and maintenance standards. These standards are updated and changed periodically as technology evolves, but this process is in place is to help ensure a safe and reliable air transportation system for the flying public.

As with much of UAS standards and regulations, deference is given to the long-proven manned aircraft standards where applicable and practical. For instance, unmanned air vehicle pilots will be required to hold pilot credentials and a medical certificate appropriate to the capabilities of the vehicle they will be operating in addition to vehicle- or platform-specific training. Maintainer standards are expected to follow a similar pattern but modified according to the size and performance of the vehicle with more stringent standards for higher performing aircraft.

Air vehicle certification standards will focus on safety, reliability, and redundancy as have their manned equivalents through the years. It is understood that it will be the spirit, rather than the letter, of manned aircraft standards that will drive UAS platform certification since there is a wider range of performance variation among UASs than manned aircraft ranging from small micro air vehicles to very large, high-flying platforms such as the Global Hawk. This fact, coupled with the fact that there are no people on board to protect, will necessitate a different approach to UAS aircraft certification than in the past.

12.4 CAREER OPPORTUNITIES

UAS career opportunities in the future will abound as the vehicles become more numerous and as the airspace opens to routine operations. There will be opportunities for UAS pilots, sensor operators, and technicians (aircraft maintenance, electronic, and information technology). Larger vehicles with a larger logistical footprint

FIGURE 12.2 General Atomics Predator™.

will require more job specialization, whereas smaller vehicle operation is more apt to require the operator to perform multiple functions such as launch and recovery, flying, maintenance, and mission payload and sensor operations.

As an example, General Atomics's Predator™ flights require a flight crew of two: one vehicle pilot and one person for sensor operations. In addition, there are separate support personnel required for vehicle launch and recovery as well as for maintenance and electronics support (Figure 12.2). Further, line-of-sight observation is required, at times requiring a manned aircraft to fly chase. For any number of smaller operators, some of these functions are combined depending on the size and performance of the vehicle. Those desiring a career in UASs would do well to select training that exposes them to a wide range of platforms and automation control software in addition to an education that exposes them to the larger challenges faced by the industry to include political and economic challenges.

12.5 AIR VEHICLE EVOLUTION

There are numerous trends to look for in the coming years related to UAS. Following are a few of the more notable.

12.5.1 Miniaturization

The size of many platforms will become smaller driven mostly by advances in materials and processing technology. Each evolution of electronics technology allows designers to build in more capability into smaller spaces. The limiting factor in miniaturization is often heat dissipation of the energy being released as more processing is being accomplished in these small spaces. In the future, as this problem is solved,

FIGURE 12.3 Microchip.

it is likely that all of the components necessary for vehicle operation in the NAS (navigation, communication, position reporting, etc.) will be located on one small printed circuit board that could be easily be removed and placed in another vehicle (Figure 12.3). In the future micro air vehicles (wingspans as small as 6 inches) and nano air vehicles (NAVs; wingspans as small as 3 inches) will become more prevalent as miniaturization technology enables.

12.5.2 Power Solutions

Energy to power UASs in the future will be the subject of much research. The requirements to become more ecologically friendly, less expensive, and more capable will stretch the limits of current sources of power into future solutions.

12.5.2.1 Alternative Energy

UASs will be no exception to the move away from fossil-based fuels, and much work has already been accomplished in this area. BlueBird Aero Systems and Horizon Fuel Cell Technologies have already fielded the Boomerang UAS powered by a 2-kg hydrogen fuel cell pushing endurance to over 9 hours. Several biofuels have been tested on UAS but it remains to be seen what role current biofuel technology will play in supplying our future energy needs. Numerous solar powered UAS have been tested with varying degrees of success. Current limitations revolve around limited payloads and number of solar arrays required to develop sufficient power along with battery weight penalties. Efficiencies here will allow researchers to translate more solar energy using less space and storing that energy in lighter, more efficient ways.

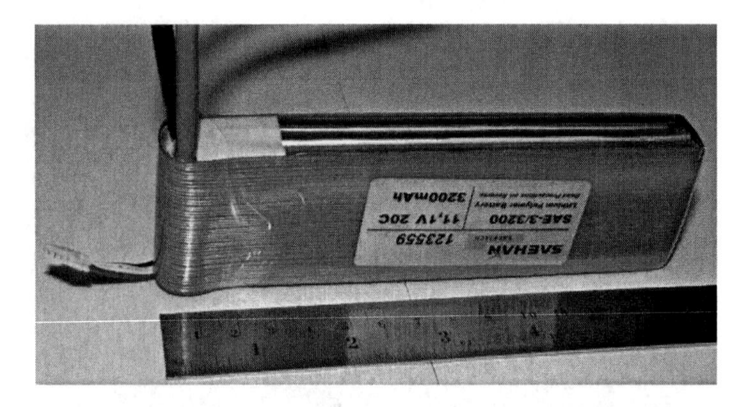

FIGURE 12.4 LiPo battery.

12.5.2.2 Electric Options

Currently, electrically powered UASs are battery-powered vehicles capable of carrying small payloads and are limited to an endurance of 1 to 2 hours at most with the weight of the battery being the largest limiting factor. Advances in lithium polymer battery technology hold much promise for extended battery life, lightest weight, and shape ability, which allows the battery to conform to aircraft design (Figure 12.4). Future advances in electric UASs will involve replenishment ability from power lines, an electric fuel "tanker" concept, or, as antennae technology develops, the transmission of electricity through the air to recharge onboard batteries.

12.5.2.3 Materials Improvements

It is axiomatic in the field of aircraft design that the less weight required for the structure of the aircraft, the more payload it can carry. Advances in structural materials will focus in large part on composite technology and will no doubt become lighter and more durable, as well as easier to manufacture, maintain, and repair (Figure 12.5). Costs are sure to escalate accordingly; however, the prices for current composite materials may correspondingly go down. Some current limitations of composite aircraft structures include long-term structural integrity especially when exposed to abnormal conditions such as in a contaminated or caustic environment. However, advances in nondestructive testing (NDT) technology should offset this limitation.

12.6 FUTURE CONCEPTS: U.S. MILITARY

12.6.1 Unmanned Combat Air Vehicle (UCAV)

The concept behind the unmanned combat air vehicle (UCAV) is to design an offensive unmanned aerial weapons delivery platform as opposed to mounting weapons on a platform that was designed for another purpose. Several designs currently exist such as the Boeing X-45A (Figure 12.6), the Northrop Grumman X-47B, and the BAE Systems Taranis. Removing the operator from on board is somewhat controversial with proponents advocating that human limits on acceleration combined with

FIGURE 12.5 Composite material.

the weight penalty for onboard life-support systems produces a vehicle with less than optimum performance. Detractors argue that computer logic will never adequately replace the human decision-making process, especially the ability to make split-second, high-consequence decisions.

12.6.2 Gorgon Stare

The Gorgon stare is a concept being fielded on the General Atomics Reaper fleet that will allow multiple end users of airborne optic data to choose from up to 12 different camera angles for a given geographic area (Figure 12.7). This will allow one aircraft to essentially provide the platform for, in essence, 12 different camera views so multiple targets can be tracked simultaneously. No doubt this technology will eventually find its way into the civilian sector.

FIGURE 12.6 UCAV.

FIGURE 12.7 General Atomics Reaper.

12.6.3 Commonality and Scalability

Given the proliferation of UAS technology and capability it is the desire of UAS operators to move toward technical commonality, which will allow for efficiencies in acquisition, support equipment, training, servicing, and support. As an analogy, some manned aircraft operators choose to operate one common model of aircraft with different variants in order to increase the familiarity of their aircraft operators, servicers, and maintainers with the equipment; much of the training and many of the features become similar allowing those who work with the aircraft to become much more efficient in performing their jobs. Likewise, the military is seeking similar system commonality for many of the same reasons. It is inefficient to attempt to field a multitude of unrelated vehicles. The concept of scalability is for the features of one vehicle to be "up-" or "downsized" based on the mission requirements and is closely related to the concept of commonality.

12.6.4 Swarming

The concept of swarming is largely a military concept (borrowed from nature) whereby a target is attacked from multiple directions, through varied means, simultaneously. It is a technique used to overwhelm a target and subdue it quickly. This concept, which is already being discussed in military circles, would involve the close coordination of multiple independent systems in a relatively small amount of airspace. In other words, these systems would need to display a high degree of interoperability most likely coupled with a high degree of autonomy in the future. The command and control infrastructure has yet to progress to a point where swarming could be supported but a move toward this concept will no doubt drive those necessary technical improvements.

12.7 FIVE YEARS AND BEYOND

This is where the discussion gets a bit more difficult and devoid of specifics since many future concepts are based on technology yet to be invented. What is known is that scientific and technological advancements are truly mind-boggling when it comes to the limits of artificial intelligence. Robots are being designed that can learn to accomplish complex tasks on their own and also learn from each other. These complex tasks involve interacting with humans, learning to speak, and generating ideas. Combined with advances in mechanics, structures and materials, and power delivery, the future is anyone's guess. Some futurists have suggested that we will see a future where artificially intelligent machines are able to repair or replicate themselves, seek their own fuel source, and make decisions that could run counter to their originally intended design. Certainly we are some distance away yet from that scenario, but if the current pace and direction of progress is taken into account we must consider those possibilities and we must also consider what we term "progress" and what is regressive in nature.

Another concept sure to continue into the future is the field of unmanned spaceflight. Certainly many unmanned space missions over the last 40-plus years have demonstrated the advantages of being able to explore space and other planets without having to consider the limitations of human physiology. The cost savings alone will spur more development in this area.

There is much more that could be discussed on this topic and certainly much more depth that could be explored on each topic, but due to the scope of this text we shall leave that for further exploration by the reader. One thing for sure, as has been said, the future is unlimited and unmanned!

DISCUSSION QUESTIONS

12.1 List the advantages and disadvantages of the term *unmanned aerial system*. What are some alternate terms?

12.2 Refer to the eight basic segments of the UAS market in Section 12.2. Use the Internet to list one current development (within the last 90 days) in each area.

12.3 List three challenges of converting ground-based infrastructure (i.e., airports) to joint manned–unmanned use.

12.4 Many UASs are designed to be used for surveillance. What challenges may arise with widespread UAS use for surveillance?

12.5 What should be the limits of artificial intelligence as it relates to autonomous decision making by UASs?

REFERENCES

Aboulafia, R. 2010, May 11. The last healthy part of the world economy: Aviation industry overview and forecast. AIA Communications Council Meeting, Arlington, VA.

Deptula, D. 2010, April 7. Remotely piloted aircraft in the United States Air Force. Keynote speech, Academic Opportunities: Developing the Future of UAS/RPA. Mississippi State University, Starkville MS.

U.S. Military Unmanned Aerial Vehicles (UAV) Market Forecast 2010–2015. n.d. Market Research Media, http://www.marketresearchmedia.com/2010/04/09/unmanned-aerial-vehicles-uav-market/ (accessed May 28, 2010).

FURTHER READING

Singer, P. W. 2009. *Wired for War: The Robotics Revolution and Conflict in the 21st Century*. New York: Penguin Press.

Appendix

The following charts are excerpts from MIL-STD-882D/E.

Example Mishap Probability Levels

Description*	Level	Specific Individual Item	Fleet or Inventory**
Frequent	A	Likely to occur often in the life of an item, with a probability of occurrence greater than 10^{-1} in that life.	Continuously experienced.
Probable	B	Likely to occur several times in the life of an item, with a probability of occurrence less than 10^{-1} but greater than 10^{-2} in that life.	Will occur frequently.
Occasional	C	Possible to occur some time in the life of an item, with a probability of occurrence less than 10^{-2} but greater than 10^{-3} in that life.	Will occur several times.
Remote	D	Unlikely but possible to occur in the life of an item, with a probability of occurrence less than 10^{-3} but greater than 10^{-6} in that life.	Unlikely, but can reasonably be expected to occur.
Improbable	E	So unlikely, it can be assumed occurrence may not be experienced, with a probability of occurrence less than 10^{-6} in that life.	Unlikely to occur, but possible.

* Definitions of descriptive words may have to be modified based on quantity of items involved.

** The expected size of the fleet or inventory should be defined prior to accomplishing an assessment of the system.

Example Mishap Severity Categories

Description	Category	Environmental, Safety, and Health Result Criteria
Catastrophic	I	Could result in death, permanent total disability, loss exceeding $1M, or irreversible severe environmental damage that violates law or regulation.
Critical	II	Could result in permanent partial disability, injuries, or occupational illness that may result in hospitalization of at least three personnel, loss exceeding $200K but less than $1M, or reversible environmental damage causing a violation of law or regulation.
Marginal	III	Could result in injury or occupational illness resulting in one or more lost work days(s), loss exceeding $20K but less than $200K, or mitigatable environmental damage without violation of law or regulation where restoration activities can be accomplished.
Negligible	IV	Could result in injury or illness not resulting in a lost work day, loss exceeding $2K but less than $10K, or minimal environmental damage not violating law or regulation.

Example Mishap Risk Assessment Matrix (MRAM)

	SEVERITY			
PROBABILITY	**Catastrophic**	**Critical**	**Marginal**	**Negligible**
Frequent	1	3	7	13
Probable	2	5	9	16
Occasional	4	6	11	18
Remote	8	10	14	19
Improbable	12	15	17	20
Designed out	21	22	23	24

Index

B

C

E

F

G

M

Q

R

T

U

V

W

Y